The Cambridge Manuals of Science and
Literature

# THE FLEA

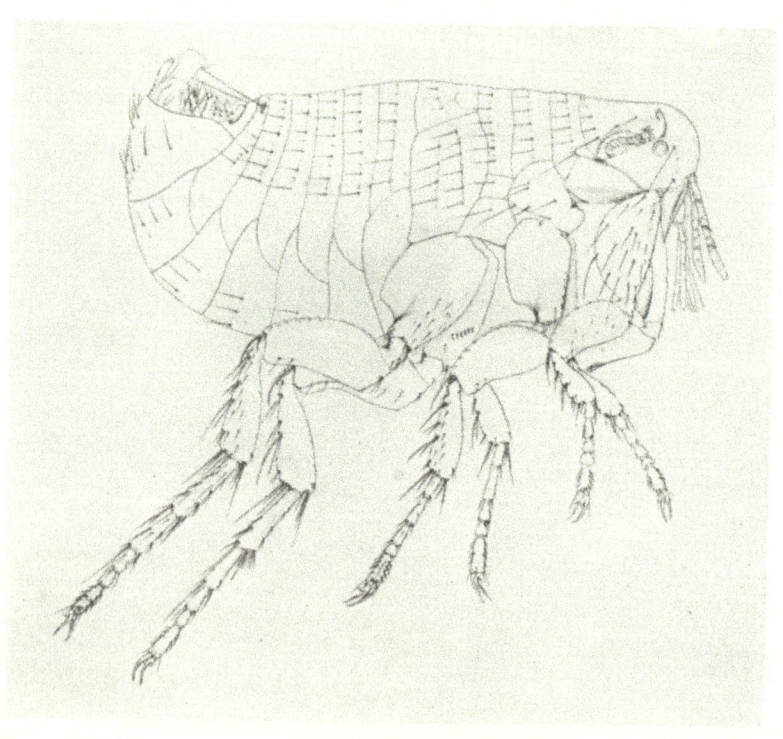

*After a drawing by Dr Jordan*

Oriental rat-flea (*Xenopsylla cheopis* Rothsch.). Male.

# THE FLEA

BY

HAROLD RUSSELL,
B.A., F.Z.S., M.B.O.U.

With nine illustrations

Cambridge:
at the University Press

1913

CAMBRIDGE UNIVERSITY PRESS
Cambridge, New York, Melbourne, Madrid, Cape Town,
Singapore, São Paulo, Delhi, Tokyo, Mexico City

Cambridge University Press
The Edinburgh Building, Cambridge CB2 8RU, UK

Published in the United States of America by Cambridge University Press, New York

www.cambridge.org
Information on this title: www.cambridge.org/9781107605831

© Cambridge University Press 1913

First published 1913
First paperback edition 2011

*A catalogue record for this publication is available from the British library*

ISBN 978-1-107-60583-1 Paperback

*With the exception of the coat of arms
at the foot, the design on the title page is a
reproduction of one used by the earliest known
Cambridge printer, John Siberch, 1521*

# PREFACE

THE aim of this book is to give in plain language some account of a small, but noteworthy, group of insects. I have avoided, whenever I could, using the technical terms of zoology. To avoid doing so entirely is impossible in a book which describes insects in some detail. No technical term has, I hope, been used without an explanation.

Over thirty years have elapsed since Taschenberg's German book, *Die Flöhe*, appeared. Our knowledge has made enormous strides since then. More species of flea are now known from the British Islands alone than were then known from the whole world. So far as I am aware, no book, devoted to what is known about fleas, has ever been published in English. The statements about these insects in the general text-books of entomology are frequently antiquated and inaccurate. But there is a fairly extensive literature on the *Siphonaptera* scattered through scientific periodicals mostly in English, German, Italian, Dutch and Russian. I have given some references in the Bibliography.

The naturalists now living who have devoted any time to the special study of fleas may almost be counted on one's fingers. In England there are Mr Charles Rothschild and Dr Jordan; in the Shetland Islands, the Rev. James Waterston; in Germany, Taschenberg of Halle and Dampf of Königsberg; in Russia, Wagner of Kieff; in Holland, Oudemans of Arnhem; in Italy, Tiraboschi of Rome; in the United States, Carl Baker and a few others. I have not mentioned medical men who have investigated fleas in connection with plague.

There are small collections of fleas in the Natural History Museums at South Kensington (London), Paris, Berlin, Königsberg, Vienna, Budapest, S. Petersburg and Washington. Of private collections Mr Charles Rothschild's at Tring is by far the best in the world. It contains something like a hundred thousand specimens and is most admirably kept. I must express profound and sincere gratitude to Mr Rothschild for having helped me in numberless ways and advised me in many difficulties.

It is well known that the mere mention of fleas is not only considered a subject for merriment, but in some people produces, by subjective suggestion, violent irritation of the skin. The scientific study of fleas has, however, received a great impetus since it has been ascertained that they are the active agents

in spreading plague.    Rat-fleas are of various kinds,
and not all fleas will bite man.    A knowledge of the
different species has suddenly become useful.    The
humble, but ridiculous, systematist with his glass
tubes of alcohol for collecting fleas, his microscopic
distinctions, and Latin nomenclature has become a
benefactor of humanity.    Some people seem to be
practically immune to the bites of fleas, but even to
such persons their visits are unwelcome.    A famous
Frenchwoman once declared : " *Quant à moi ce n'est
pas la morsure, c'est la promenade.*"

H. R.

LONDON,
   *September*, 1913.

# CONTENTS

# LIST OF ILLUSTRATIONS

# CHAPTER I

FLEAS form a group of insects that have, until recently, been little studied by zoologists. We call them insects because they are jointed animals, or Arthropods, with three pairs of legs in the adult condition. The reader will best understand the position which fleas occupy in the general classification of animals by remembering that the arthropods, or jointed animals, are one of a dozen subkingdoms, or phyla, to which the various members of the great animal kingdom have been assigned. There is good ground for believing that all the animals included in each phylum trace their ancestry back to a common primitive form which lived in more or less remote ages. Besides (1) *Insects*, the arthropods, or jointed animals, include (2) *Crustaceans*, such as crabs, lobsters, shrimps, wood-lice, water-fleas and barnacles; (3) *Myriapods*, such as centipedes and millipedes; and (4) *Arachnids*, such as spiders, scorpions, mites and ticks. To all these varied forms of animal life fleas, and other insects, are therefore more or less nearly related.

The animals belonging to this large and important collection, which compose the arthropod phylum, have certain common characteristic features. We find a body made up of a series of more or less completely similar segments placed one behind the other. In this they resemble certain worms which are far less highly organised. The body is elongated, symmetrical on either side, and the mouth and anus are at opposite ends. There is, however, an important advance on the segmented worms. Each typical segment carries a pair of appendages which are very different from the foot-stumps that are found on certain worms. These appendages of arthropods are divisible into distinct limb-segments, separated from one another by moveable joints, and acted upon by special muscles.

The common ancestor of all the various arthropods which are found living on the earth to-day, was probably composed of a series of segments each very similar to the last and each bearing a pair of very similar appendages. In the course of ages, these appendages have been astoundingly modified in form and in function. So it happens that we find in the arthropods of the present day pairs of antennæ, of mandibles and other mouth-parts, of pincers, of legs, of swimming-feet and of tail pieces which on close examination can all be traced back to a common structure. The body-segments, also, have been strangely fused together and modified. All that has been so far said applies equally to fleas and to other insects.

It is of great interest, when one comes to make a minute study of the form and external structure of a flea, to try and trace the modifications that must have taken place in the course of descent from the ancestral arthropod; but the relationship of fleas to other insects living at the present day is of more immediate concern. Insects are highly specialized arthropods and fleas are highly specialized insects. This means that they have become vastly modified from the primitive ancestral type and fitted thereby for a life among certain defined and peculiar surroundings.

It will be unnecessary to remind the reader who knows anything of zoology or of botany that all classification is now based on descent. Since naturalists have abandoned a belief in the special creation of the various species of animals now living on the earth and have conclusively shown that they have arisen by descent and modification from other forms, the problem is to reconstruct a vast genealogical tree. What then were the ancestors of the fleas and to what other insects, in consequence, do they appear to be related?

It is probable that the ancestors of the fleas were winged insects, and that the organs of flight were

gradually lost, as they became useless, when a
partially parasitic life was adopted. At one time
entomologists regarded fleas as wingless flies and
placed them in the order Diptera. Certain supposed
scaly plates on their bodies were regarded as the
atrophied relics of wings. It is, however, more than
doubtful whether this view is correct; and all modern
entomologists who have given any special study to
fleas are agreed that they are sufficiently unlike any
other living insects to deserve a place in an order by
themselves. To this order the name *Siphonaptera*
has been given: which means that the insects com-
prised in it are provided with sucking mouths and
are destitute of wings. Another name for the order
is Aphaniptera, but this is gradually falling into
disuse. Linnæus (1758) only mentions two species
of flea: the human flea which he appropriately named
*Pulex irritans,* and the chigoe of hot countries which
he called *Pulex penetrans,* from the habit which the
female has of burrowing under the skin of her victims.
At the time of writing, about 460 species of flea have
been described and named; but some of the names
are doubtless synonymous, and the actual number of
separable species that have been discovered is some-
where about four hundred. The vast majority of
these have been described within the last few years,
which shows what can be done when attention is
turned to any neglected group of animals. There

can be no doubt that many undiscovered species still remain, and will now, in due course, be collected, described and named.

The position which should be assigned to the order Siphonaptera in the general scheme of insect classification is a question on which the most learned modern entomologists have disputed with considerable vigour. Some see the nearest relatives among the beetles, others among the flies. The majority, as we shall see later on, would place them near the Diptera: but since no convincing arguments have been produced on either side it may be wisest to regard the question as still at present unsolved.

Fleas belong to one of the groups of insects which go through a complete metamorphosis. Their life-history consequently falls into four divisions: egg, larva, pupa and imago. If the climate permits, the female flea lays her eggs all the year round, and from one to five are dropped at a time. Unlike those of many other parasites they are never attached to the hairs of the hosts, but appear to be deposited indiscriminately on the floors of houses or in the nests and sleeping places of their hosts. The eggs generally hatch in a few days, and a minute, white, wormlike larva emerges (Fig. 1). The larvæ of some, and possibly of all, fleas are provided with a wonderful adaptation in the shape of an egg-breaker or hatching-spine. This is a thin plate, like the edge of a knife, where

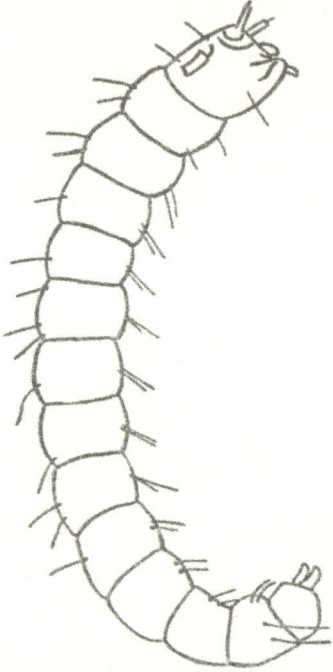

Fig. 1. The larva of a flea. The body consists of thirteen
segments and is legless. On the fore part of the head
are the antennæ and on the upper part of the head is
shown the knife-like edge of the egg-breaker. The mouth-
parts are adapted for biting. On the last segment of
the body are the two caudal stylets.

the point of the head comes in contact with the shell.
The movements of the prisoner make a slight split in
the egg-shell, which then bursts asunder. This organ
has vanished in later larval life, and it is probably
lost after the first moult. The larva is legless and
has thirteen segments. It grows rapidly, and, as it
grows, moults its skin several times. It is provided
with mouth-parts adapted for biting, and eats any
decaying organic refuse. The larvæ may be reared on
the sweepings of an ordinary room or the dirty scurf
which collects at the bottom of old birds' nests. It
is hardly necessary to add that the mother takes no
interest whatever in the larvæ and that the belief
that she feeds them on dried blood is not based on
any sound foundations.

The larval stage lasts some days, and the animal
spins a small cocoon before pupating. In the course
of a few more days, the time probably depending on
the weather, the perfect flea emerges. The larvæ
generally live in places where the perfect insects will
have an opportunity of finding a host as soon as
they leave the pupal envelope. The nests of their
hosts where the young are being reared are always
favourite places. It seems possible that the com-
parative immunity from fleas which hoofed mammals
or Ungulates enjoy may be due to the fact that the
young beast follows its mother from the time of birth
instead of passing its early life helpless in a nest.

Observations made on the development of the dog-flea (*Ctenocephalus canis*) in India show that eggs laid on October 17 hatched on October 19. The larva spun its cocoon on October 25 and the mature flea emerged on November 2. In Northern Europe the human flea takes about four weeks in summer and six weeks in winter to pass through its metamorphosis.

Unlike many parasitic insects, fleas do not constantly pass their time upon the bodies of their victims. The greater part of their life is probably spent on the ground, in the house, or nest, of the mammal or bird which serves them with blood. In this respect there is considerable difference in the habits of different species of flea. Some attach themselves to an animal and actually burrow into the skin. These are the most parasitic species. Some only come to feed and leave to lay their eggs. Many probably do not suck blood more than once in their lives.

An animal which harbours fleas and which nourishes the adult insect with blood is called a *host*. No fleas are more than what is called temporary parasites; which means that they pass but a portion of their lives on their hosts and frequently take occasion to hop on and off. All fleas, apparently, go from host to host. The labours of diligent collectors have proved that the great majority of mammals and birds have fleas. As a general rule, it is true to say that certain species of flea are associated with certain species of host. Thus man is the true host of *Pulex irritans*; the cat family are the true hosts of the cat-flea (*Ctenocephalus felis*); and the dog family are the true hosts of the dog-flea (*Ctenocephalus canis*). But the human flea is sometimes found on cats and dogs, and cat and dog-fleas occasionally bite human beings; and cat-fleas are found on dogs and dog-fleas are found on cats. All fleas, so far as we know, may occasionally pass from one species of host to another; but they do not, for the most part, seem to flourish in unaccustomed quarters. Some fleas are more catholic in their tastes than others. Some seem to be very strictly confined to one host, and even when starving only suck strange blood under protest. There is a species of flea that has only (except by accident) been found on the long-tailed field-mouse and another that has only been found on the hedge-hog. Other fleas are commonly found on two absolutely distinct animals; a good instance of this is the human flea which, at all events in certain parts of England, is a regular parasite of the badger.

As distinguished from true or natural hosts one must separate what may be termed casual or accidental hosts. All animals which come in contact with one another, or which live in close proximity, may exchange fleas. So even bird-fleas may be collected from mammals and typically mammalian fleas from birds. In this fashion puzzles may arise which tax the ingenuity of the collector to solve. Bird-fleas are sometimes found on bats, and this may be obviously attributed to the bats having inhabited a hole which was tenanted by starlings or an old loft infested with the fleas of pigeons. All beasts of prey are sometimes found to harbour the fleas of animals they have devoured. Rabbits' fleas are found on wild-cats; hedgehogs' fleas on foxes; mice fleas on weasels; and fleas characteristic of small birds on stoats. So also in the case of mice, rats and voles with holes and runs in the same hedgerow, the parasites usually peculiar to one are not uncommonly found on the others. It is sometimes difficult to determine the true host of a flea.

Much more puzzling to explain are the reasons which confine a flea to a certain host and which cause closely allied hosts to have different fleas. The fleas from the house-martin and the sand-martin are quite different; those from the domestic fowl and the domestic pigeon are distinct species. The causes which have affected the evolution of the various forms of flea are too obscure to enable anyone at the present day to offer any satisfactory explanation.

Speaking generally, the fleas found on birds have points in common, and they probably form a natural group to themselves. What may be called true bird-fleas have been collected from almost all European birds. An unwieldy genus (*Ceratophyllus*) comprises many species of different flea. Some species are very abundant and infest the nests of many different birds. Others are extremely rare. One of these rarities (*C. vagabundus*) is found in the nests of puffins and other sea-birds. Another has been collected on antarctic petrels. Penguins have a special genus of flea to themselves. A specimen, unique at one time (*Ceratophyllus borealis*), in Mr N. C. Rothschild's collection was obtained from the gannet. It has now been found on rock-pipits in the Shetland Islands.

Two very rare fleas (*C. farreni and C. roth-schildi*) are found in the nests of house-martins; yet the nests of these birds are infested with common species besides. A plague flea (*Xenopsylla*) has been found on an African swift.

Forty-six different species of flea have been found in the British Islands, but many of these are extremely scarce.

We know too little about the geographical distribution of fleas to lay down many accurate generalities. When a great deal more material has been collected and studied, it may be possible to show that certain groups are associated with certain regions of the earth or certain orders of animals. To some extent this is already seen to be the case. The fleas

indigenous to the New World are distantly related
to those of the Old World. Broadly speaking the
geographical distribution of the parasite must follow
that of the host. But sometimes the parasite is
impatient of cold and cannot follow the host out
of the tropics. The chigoes and their allies are fleas
of hot countries. Different kinds of bats are found
from the tropics to the Arctic circle, but the same
bat-fleas are not found everywhere.

When a flea has a cosmopolitan range it is
probable that it has travelled over the world in
company with its host.

Monkeys have no fleas. This is an assertion that
is commonly received with surprise and incredulity.
Occasionally a gorilla or a chimpanzee may get a
chigoe in its toe. And monkeys in zoological gardens
or menageries are possibly exposed to the danger
of catching an occasional human flea from the people
who crowd round their cages. These are remote
contingencies which may happen to anyone. Healthy
wild monkeys are much too clean and active to
harbour fleas. When they search one another's fur
in a fashion that must be familiar to most persons,
they are clearing their coats of particles of scurf or
of similar scraps of dirt and not of fleas. So,
speaking generally, it may be said that no fleas have
been found truly parasitic on monkeys.

Bats have fleas, but not in great abundance. All

bat-fleas are rare on their hosts and extremely difficult to find and collect. The same species are not found on fruit-bats and the ordinary smaller insect-eating bats. The geographical distribution of some bat-fleas is puzzling. For instance, one species (*Ischnopsylla unipectinata*) is found on the greater horse-shoe bat in Europe; but it is, apparently, not found on the same bat in the British Islands. In Somaliland and in India it is found on other bats.

With certain exceptions, Ungulates are remarkably free from fleas. This great order of mammals includes a variety of hoofed animals: oxen, sheep, goats, deer, pigs, camels, giraffes and antelopes.

The only true fleas found on these are two species of the genus *Vermipsylla*, which resemble the chigoes in so far that the pregnant females burrow into the host and expand there. One species has been found on camels and horses in Transcaucasia; another on roe-deer in Northern China. The female of the last is often found ensconced on the inside of the nostrils of the deer. Of course chigoes may attack domestic Ungulates of all kinds; but no other members of the family *Pulicidæ* or typical fleas except those two above mentioned have been found on hoofed mammals.

Insectivora such as moles, shrews and hedgehogs are the hosts of a great variety of species. The same thing may be said of the Rodents, which include

porcupines, squirrels, rats, mice and a vast number
of other small mammals whose geographical distribution includes almost the whole of the habitable
globe.   Probably more different species of fleas have
been collected from Insectivora and Rodents than from
all the other orders of mammals grouped together.

The Carnivora, excluding the Pinnepedia, or seals,
sea-lions and walruses, harbour numerous species.

Among the Edentata a very remarkable and highly
specialised genus of fleas is parasitic on armadilloes
in South America.    This genus (*Malacopsylla*) consists of two species only, which are confined to South
America and are found on the armadilloes and on
carnivorous animals which probably have preyed on
them.    The thorax of these fleas is much reduced
and very small in size.    Their piercing organs are
slender and weak, but they possess enormous spines
on the legs with which they hold on to their hosts.
These two South American fleas (*M. grossiventris*
and *M. androcli*) will be referred to again later as
striking examples of fleas with strongly developed
legs and weakly constructed mouth-parts.   The con-
trary combination of powerful mouths and degenerate
legs is also found in other groups of fleas, as will be
seen in the chapter on the chigoes.

The Marsupials of Australia and South America
have special fleas which were probably associated
with this strange order of pouched mammals before

they became divided into the American and Australian groups. Fleas have been collected on the spiny ant-eater (*Echidna*) which belongs to the lowest order of Monotremes or egg-laying mammals.

On almost every form of bird, including the most aquatic kinds, fleas of various species have been obtained.

Only one instance has been recorded of a flea occurring on a reptile. A female of one of the species of burrowing chigoes (*Echidnophaga ambulans*) from Australia was collected by Dr Woodward from the Brown Snake (*Diemenia superciliosa*). This reptile, which is well known in Australia, belongs to a sub-family that contains some of the most deadly poisonous snakes and is allied to the cobras. The Brown Snake is a terrestrial snake, and one must regard the presence of the flea on such a host as a rare and chance occurrence. The snake was captured at Herdman's Lake, near Perth in West Australia. The same species of flea has also been obtained from the phalangers (*Trichosurus*) which live in the tops of the Australian gum-trees; from the little terrestrial and nocturnal rat kangaroos (*Bettongia*); and from the banded ant-eater (*Myrmecobius*), another Australian Marsupial. It is possible that the flea moved from some small mammal which was being devoured by the snake and managed to fix itself between the scaly plates of the reptile.

When fleas are hatched in a nest they have no choice but to attach themselves to the young mammals or birds. But even in that case they frequently leave their hosts and do not for very long remain stationary. Moreover, when a host dies and becomes cold the fleas invariably leave their quarters, which explains how it may happen that Carnivora get infested with the fleas of their prey. This change of hosts which is always occurring makes it impossible to draw conclusions from material collected in zoological gardens where many animals are herded together. In menageries, too, the normal conditions of breeding are absent. A German naturalist collected 2036 fleas from theatres, concert-halls, ball-rooms, schools and barracks in the grand-duchy of Baden and found that more than fifty per cent. were dog-fleas (*Ctenocephalus canis*). What the proportion may be in other parts of Europe we have no materials from which to form a judgment. In zoological gardens cat-fleas (*Ct. felis*) are generally numerous in most of the cages.

It is, of course, well known to every zoologist that species are not fixed or constant and that various forms of mammal or of bird tend to show geographical variations. When a long series of skins are laid out on a table and carefully examined it is seldom that those from the west of any great region cannot be picked out and distinguished from those obtained in

the east. So we also get northern and southern forms
of the same species varying slightly. These variations
are perceptible in many forms of insects, and zoologists
now describe these local races as subspecies and
designate them with trinomials. No one, however,
knows enough as yet about all the various forms
which are assumed by fleas to attempt, except in a
few instances, to do so in the case of these animals.

The study of Siphonaptera is still quite in its
infancy. We know little or nothing of the minute
geographical variation of fleas. That there is such
a thing can already be seen in a few species. In the
meantime the study of variation must be postponed
until collectors have amassed a more plentiful amount
of material; and it is best to treat all forms which
are to all appearance constantly different as being
specifically distinct until more is known about
variation.

Any classification of fleas that may now be
attempted can only be tentative. It will be enough
for present purposes if the reader will remember
that the Order Siphonaptera can be divided into
three groups or families: (1) the chigoes and their
allies, which are the most parasitic fleas (*Sarcopsyllidæ*); (2) the typical fleas to which the majority
of species belong (*Pulicidæ*); and (3) the bat-fleas
(*Ceratopsyllidæ*), which have certain peculiarities that
will be described in a later chapter.

Of the antiquity of fleas, and of the period in geological history when the order made its appearance, little can be said.   When it was thought that fleas were confined, as parasites, to warm-blooded mammals and birds, evolutionists were inclined to say that the parasites could not have appeared before their hosts.   The discovery of a flea on a reptile opened the vista of possibly enormous antiquity stretching back to Permian or Carboniferous ages. The fossil record is most meagre.   If we reject as too doubtful the supposed remains of a flea from the lower Oligocene strata at Aix in Provence, only one undoubted fossil has been discovered.   Nor does it seem certain that fleas are entirely restricted to preying on vertebrates.   Dr Dampf introduced a number of common bird-fleas (*Ceratophyllus gallinæ*) of both sexes to some hairy caterpillars.   He observed that several of the fleas buried their heads in the hairy covering of the larvæ and remained some time in the attitude of sucking blood.   While this was going on the victims made violent demonstrations of annoyance and discomfort.   He also observed that a naked caterpillar was not attacked.

Mr Boden has also recorded how he found in a seed-warehouse some peas that were being eaten by two species of Lepidopterous larvæ.   On bringing these home and keeping them in a jar, he found among them some small larvæ which ultimately turned into fleas, probably *Pulex irritans*.   These fleas, being confined without other food, were observed to prey on the Lepidopterous larvæ and to feed freely on their juices.   The larvæ which were attacked by fleas pined and died.   The fluid from the stomach of the fleas when they were crushed was transparent and not red like vertebrate blood which often exudes when a mammalian flea is pinched and cracked open.

A French entomologist has also reported that the numerous fleas which swarm in the dwelling-houses of Corsica, for want of other nourishment turn their attention to flies that may be incapable of flight.

The only fossil remains of a flea that have, so far, been discovered are a single insect in a piece of Baltic amber of Oligocene age.   Many organic remains have been preserved in this fashion, but this is the first mammalian parasite that has been found.   The flea is admirably protected by its semi-transparent sur-roundings, and the most minute details of structure, the arrangement of bristles on the body, and the number of segments to the labial palpi can be dis-cerned.   This unique object is in the collection of Professor Klebs.   The first point to note is that a flea of this antiquity hardly differs from the existing insects of the present day.   It has been referred to an existing genus (*Palæopsylla*) of which there are

at least four species living.   Three of these are parasites of the mole, and the fourth is found on shrews.
There is good reason to suppose that the host of the
fossil was some insectivorous mammal.   The early
specialisation of fleas is strikingly illustrated.   This
insect is already adapted for life on some warm-
blooded animal.   It has a thoracic comb, and its
mouth-parts are in all respects like those of a modern
flea.   It belongs to a genus which is still commonly
distributed over Europe.   When we consider how
remote are the chances that a mammalian flea should
first get embedded in amber and should, subsequently,
be detected and described by a naturalist, we may
well understand that the owner of the fossil asked,
though without success, £1200 for it.

The ordinary person regards fleas as a subject for
humour of an obvious and familiar kind.   The utili-
tarian despises a man who can cheerfully spend his
time in collecting fleas.   Yet it seems probable
that a study of their forms and habits may be of
immediate benefit to the human race.   The discovery
that fleas are connected with the spread of plague is
an instance of apparently unprofitable scientific labour
proving of direct advantage to mankind.   An accurate
knowledge of the structure and habits of fleas is now
seen to be of importance to all who are engaged in
fighting one of the most dreaded infectious diseases.
When plague breaks out men of science now at once

turn their attention to the fleas.   This is likely to
prove more directly efficacious than the mediæval
custom of marking the house with a red cross and
inscribing the legend, "God have mercy on us."


## CHAPTER II

### THE EXTERNAL STRUCTURE OF A FLEA

IN comparing the structure of a flea with that
of a man, or any other of the higher animals, it is
of the utmost importance to understand that the one
has an internal and the other an external skeleton.
In either case the skeleton serves as an attachment
for the muscles by which the animal moves itself.
Everyone is familiar with the external skeleton of a
lobster and can see for himself how the muscles are
attached.   The structure of a flea, though so much
smaller, is somewhat similar, except that the skeleton
is composed of a horny substance known as *chitin*
instead of being calcified.   The chitinous cuticle
entirely covers the flea, but it varies in hardness
and thickness on different parts of the body.   The
epidermis, or true skin, lies immediately beneath.
On those parts of the body which are to the rear
of the head the chitin forms a series of plates or
shields which overlap one another somewhat like the

tiles of a house.   The segmented structure of a flea
is there most clearly seen; this we may suppose is
an inheritance from the segmented worms.

The chitin which forms the external skeleton of a
flea is secreted by an outer layer of cells on the
insect's body.   The deposit being thin at the joints,
and thick on the plates, which serve for protection,
the flea is encased in a suit of flexible armour.   It
is made of a fairly solid and dense substance, but,
owing to the absence of carbonate and phosphate of
lime, is much lighter than the familiar external
skeleton of the lobster.   Chitin is a very peculiar
and durable substance which resists boiling in acids
or alkalies.   It is a structureless substance, in the
sense that it does not consist of cells.   Though horny
in appearance it is, of course, in no sense true horn
like that of the nails, hoofs, claws, and horns of
vertebrates.

The different species of flea vary considerably in
size.   Some are smaller than the familiar human
flea.   Others are much larger.   A very large flea
(*Hystrichopsylla talpœ*) is that found on the mole.
The largest known flea (*Dolichopsyllus stylosus*) is
found on small rodents in the United States where,
as we know, all things are on a bigger scale than in
the Old World.   It is seven millimetres long.

The colour of the horny integument varies from
a pale or light yellow to a ruddy or dark brown.

It is plentifully sprinkled over with spines, bristles, or hairs, directed backwards so as not to impede progress. The presence or absence, the arrangement on the body, and the size of these serve, along with other features, to distinguish different species. They seem materially to help a flea in those wriggles to escape with which we are all familiar. The bristles are not always the same in size and arrangement in the two sexes of the same species of flea. As a rule the males are more bristly than the females.

These appendages of the flea's integument are called by various writers either spines, bristles, or hairs. There is, however, no real distinction in the structure or nature of the appendages, and it is a question of degree which name is most appropriate.

In distinguishing species, very little reliance can be placed on the colour of the flea. An insect newly emerged from the pupa is always lighter in colour; and the difference between the appearance of an empty stomach and a stomach gorged with red blood is surprising.

The general external appearance of a typical flea belonging to the main family *Pulicidœ* is fairly familiar to most persons. This is well seen in the figure of the oriental rat-flea. The body is compressed or flattened from side to side, and this is a feature which is extremely rare among insects. It doubtless enables the animal to glide with greater facility through the hairs of its host. Like other insects, a flea is readily divided into a head, a thorax, and an abdomen. The head is rounded on the top and front and shows no obvious trace of segmentation ; but what is known of the development of other insects leads one to think that it must properly be regarded as a number of segments closely fused together. On the under side of the front part of the head is a beak or proboscis for piercing and sucking, composed of the mouth-parts, whose structure is worthy of minute study. It will be best to examine them in detail in a subsequent chapter.

Some fleas have eyes, others have none. The common mouse-flea (*Leptopsylla musculi*) is blind. The bat-fleas are also destitute of eyes. The nocturnal habits of their hosts would render eyes of little or no use. If eyes are present they are large and placed on either side of the head. Each is a simple eye or ocellus ; the compound eyes, divided into a great number of hexagonal facets, which are characteristic of many insects, are never found in fleas.

Nothing is known about the flea's powers of vision, but there is no reason to suppose that they are at all acute. The eyes are marked with pigment. Ocelli appear to be primitive types of insect eye which are, perhaps, an inheritance from a wormlike ancestor. Presumably all the fleas of long ago had eyes and those that are now blind have lost their organs of sight from disuse. In their simplest condition, the eyes of the lower invertebrates only enable the creature, so far as one can judge, to distinguish light from darkness. Entomologists believe that the power of vision of ocelli is probably confined to very near objects and that this simple form of eye is more useful in dark places than the compound eyes. There is no reason for believing that fleas can distinguish colours or can discern any object which is more than a few inches away. It is enough for their purpose to perceive from which point light comes upon them and to make all despatch to escape in the opposite direction.

In blind fleas there is often a spine where the eye should be. In one species the spine is rudimentary and there is some black pigment beneath it. It is not impossible that this is the vestige of a once functional eye. In one genus, however, the eye and the spine are both present. Of the fleas belonging to this genus one species is South American and the other European. The latter (*Typhloceras poppei*) is confined to the long-tailed field-mouse.

The organs by which fleas keep in touch with the outward world, and with other fleas, are their antennæ. All fleas have antennæ; but unlike those of a moth, a beetle, or a grasshopper, each fits neatly into a groove at the side of the head and can be protruded when desired. This is another adaptation to enable the insect to creep swiftly through a forest of hairs.

The combs which are found on the heads of many fleas are organs of exceptional interest (Fig. 2). They are toothed and horny appendages, which are connected with parasitic habits, for somewhat similar

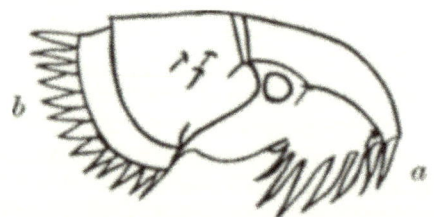

Fig. 2. Showing a type of (*a*) *genal* and (*b*) *thoracic* combs of a flea, on the under part of the head and on the thorax respectively. Analagous combs are found in several other parasitic insects and on the abdominal segments of certain fleas.

combs are found on several unrelated groups of parasitic insects, as, for instance, on parasitic beetles (*Platypsyllus*) found on the beaver, on insects allied to bed-bugs (*Polyctenes*) found on bats, and on wingless flies (*Nycteribia*) which infest Egyptian and South American bats. The majority of the *Pulicidæ* have one or more combs with comparatively long teeth. These combs reach their maximum

development in the bat-fleas which have no less
than eight.  Some Australian and South American
fleas (*Stephanocircus*) have a helmet-like comb
extending all round their heads.  These combs are
by some supposed to be of service in holding on to
the hairs of the host ; and, if one may judge from
experiments made on live fleas in cotton wool, they
are also used in moving forwards through the fur.

All the chigoes (*Sarcopsyllidæ*) have a large
triangular post-oral process which is more or less
curved and probably prevents the flea slipping back
as it pushes forward.  Bat-fleas (*Ceratopsyllidæ*)
have lobes or flaps placed two on each side of the
head, which may possibly serve an identical purpose,
but whether they do so is not known.

These combs may be divided into three groups
according to the part of the flea's body on which
they are found.  Those found on the head are called
genal combs and take the form shown in Fig. 2.
There are also combs found on the thorax.  The
fossil flea (*Palæopsylla klebsi*) described in the
previous chapter has one of these thoracic combs.
A certain number of fleas also have combs on the
abdominal segments.

There are really two types of toothed organs on
fleas to both of which the name of comb may be
given.  One is composed of a sheet of chitin with a
number of slits and teeth and resembles a true comb.

The other consists merely of a number of highly chitinised bristles arranged in a row. They probably both serve the same function. Apart from their use as organs to assist movement onwards, they may also serve as *hair-tight* joints and protect the flea from the inconvenience of getting the tips of the host's fur into the joints of its horny armour.

The size of the head compared to the thorax and abdomen varies considerably in different species. Some fleas have what may be called by comparison large heads and others very small ones.

A small head is never found in a flea with powerful mouth-parts. The head being the bearer of piercing and sucking organs, which require strong muscles, there must be room not only for the organs but for their extensors and retractors.

There are normally three rows of bristles on a flea's head which divide the head into four sections. It is possible that these correspond to the four segments of the ancestral insect which are now fused together.

The head of a flea is closely applied by the whole of its back surface to the body and that slender and conspicuous neck which is characteristic of the Diptera, or flies, is not to be found in any fleas. For this reason a flea cannot turn its head in any direction without at the same time following it round with its body.

The thorax of a flea consists of three segments called respectively the prothorax, mesothorax and metathorax. The chitinous external skeleton which covers each of these three segments is primarily a hoop but each hoop is further subdivided into a number of complicated plates. Attached to the thorax are the three pairs of legs which are characteristic of all adult insects. The hind pair are very much the strongest (Fig. 3). They are the organs of hopping. It has often been pointed out that if men had the leaping powers of some fleas they would bound with ease backwards and forwards over the cross on the top of S. Paul's Cathedral. Each leg consists of four segments beautifully articulated and plentifully supplied with bristles. At the end comes the foot with five very short segments. The last segment is provided with a pair of more or less formidable claws. Fleas use their legs for leaping, for running, and for clinging to their hosts. They also use their mouth-parts for the last purpose and it is worthy of note, as we shall see later on, that in those fleas in which the mouth-parts are shortest and weakest the legs are most liberally supplied with bristles and possess the stoutest claws. The legs of a flea are unique in the insect world owing to the enormous development of the segment nearest the body called the coxa. Most leaping insects rely for their activity on the muscles of the lower joints. In a grasshopper

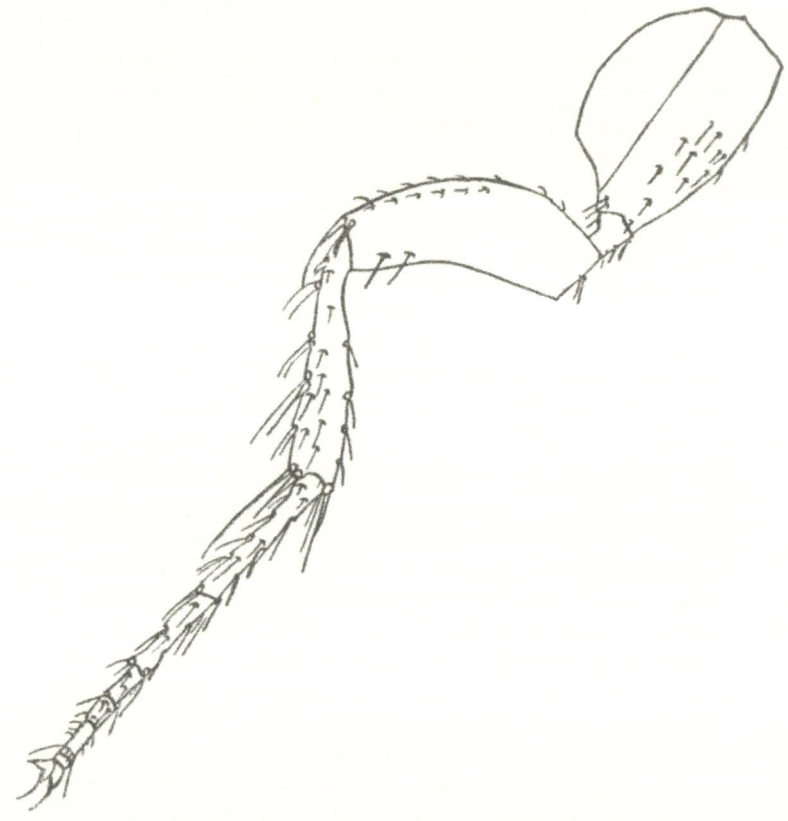

Fig. 3.   The hind leg of a flea.   The segment or joint nearest the
body is the *coxa* which is unusually developed.   Next come the
small *trochanter* and the larger *femur*.   The *tibia* which is long
and slender follows.   Then come the five *tarsi* with the sixth
and ultimate segment provided with claws

it is the third joint from the body (femur) which is so immensely enlarged. The three pairs of legs are each attached to a different thoracic segment.

When fleas walk, they are so to speak plantigrades walking on the sole of the foot, and all the tarsal or foot joints are applied to the surface of the ground. The claws serve as grips so as to make the most of any unevenness ; and thus the insect drags itself along with surprising rapidity when it moves through the hairy coat of a mammal. But on an open surface fleas are not really rapid movers compared with many other insects.

The two claws on the end of each ultimate foot segment are freely moveable and are in fact highly modified bristles or setæ.

In all fleas one of the plates of the metathorax (or hindmost thoracic segment) called the epimeron, is large and prolonged towards the rear. It invariably bears a stigma. The epimeron is placed laterally to the first abdominal tergite. The older naturalists jumped to the conclusion that this was the remains of a wing. The best judges have, however, formed a decided opinion that no trace of the relic of a flying organ can be detected on the thorax of a flea. Heymons, a German entomologist, has also failed to detect any sign in dissections which he has made of the larvæ and the pupæ.

The epimeron is in fact neither a scale nor a

wing but a portion of the thorax present in all
insects.   It is of no special service to the flea except
as a portion of the thoracic armature which covers
the body.

The larva of a flea has no legs ; the adult insect
has six.   A study of other embryo insects shows that
the ancestors of insects had many legs.   It is an
interesting problem why insects lost the legs on
their abdomens, why legs should now invariably be
restricted to the thorax, and why there should never
be more than three pairs.   In the earliest known
insects which lived on the earth, before winged forms
were evolved, the number of legs was already six.
But our knowledge of fleas is too small to attempt, at
present, to trace their exact line of ancestral descent.

The abdomen of a flea consists of ten segments.
The horny plates which cover the dorsal side are
called tergites; those on the ventral side sternites.
In fleas, as in all holometabolous insects, that is
those which pass through a complete metamorphosis,
the sternite of the first abdominal segment is suppressed and has completely disappeared.   The tergite
which covers the dorsal part of the first abdominal
segment nearest to the thorax is, however, always·
present.

The ultimate segments of the male and female
flea are modified for reproductive purposes and of
these segments more must be said later.

Having now given a rough outline of the external skeleton of a flea, it only remains to say something about the muscular system. Attached to the inside of the chitinous armature are an enormous number of muscles, whitish and almost transparent. They act as extensors, retractors, flexors, elevators and depressors. The joints and hinges of the skeleton allow of considerable, but not perfect, freedom. The muscles of locomotion are partly in the thorax and partly in the several joints of the legs. Our knowledge of the muscular system of fleas is very imperfect. But, as in other insects, the general arrangement of the muscles is based on the segmented structure of the body.

For the reader who can accurately picture to himself the external structure of a flea and of the typical insects belonging to other orders, a few words may be said on the probable ancestry of fleas and their relationship to other living insects. This vexed and much debated question is still, as the older naturalists would have said, *tremendum mysterium*. Very little light has yet been thrown upon it, and the most divergent views have been expressed by learned and competent entomologists. A historic survey of the various opinions that have been held since the days of Linnæus would fill many pages ; but a short summary of the different orders to which fleas have been referred by different zoologists will suffice.

The older authors, Linnæus, Geoffroy, Cuvier and Duméril, and Gervais placed them among the *Aptera* because they were wingless. Kircher regarded them as *Orthoptera*, an order which includes grass-hoppers and crickets ; but he has had few followers. By Fabricius and by Illiger they were treated as *Hemiptera* or bugs. Lameere, a Belgian, has recently expressed a decided view that fleas are really a family of *Coleoptera* or beetles. Those who have held the once orthodox opinion that they belonged to the *Diptera* or flies are Roesel, Oken, Straus-Dürkheim, Burmeister, Newman, Walker, von Siebold and Wagner.

The structure of an adult flea, however, differs from that of an adult fly in the following note-worthy respects : the mouth-parts are differently constructed, the head of the flea is closely joined to its thorax, the three divisions of the thorax are not joined and fused, the flea is wingless, the eyes of fleas are simple ocelli, and there are differ-ences of lesser importance in the stigmata, which give access to the tracheal system by which all insects breathe.

The number of those who have regarded fleas as belonging to a distinct order of insects is consider-able : they are Lamarck, De Geer, Latreille, Kirby and Spence, MacLeay, Leach, Dugès, Bouché, van der Hoeven, Westwood, Landois, Brauer, Kraepelin,

and Taschenberg. Modern opinion is all but un-
animous on this point.

There remains, however, a second question. Even
if it be agreed that there must be a distinct order
for *Suctoria, Aphaniptera, Siphonaptera,* or fleas ;
where ought that order to be placed ? In which
other order of insects must we look for the nearest
relations of fleas ? For a time after the acceptance
of the fact that insect forms have been evolved, and
not separately created, the ancestors of fleas were
searched for among some species of fly.

Then Kraepelin rejected the view that flies were
as closely related to fleas as most entomologists
thought and his followers could only find points of
difference and no points of resemblance. Dahl (1899),
a German, then took up the cudgels for the fly
theory. Dahl pointed out the resemblance between
fleas and a group of flies called *Phoridæ* also
parasitic on warm-blooded animals. During the
ensuing years the debate was resumed afresh
with much liveliness and sometimes with a little
acrimony.

The fleas were placed by MacLeay and by
Balbiani between the *Diptera* and *Hemiptera* ;
by Leach between the *Hemiptera* and *Lepidoptera* ;
by Dugès between the *Hymenoptera* and *Diptera* ;
by Brauer between the *Diptera* and *Coleoptera.*
Handlirsch thinks that fleas have no connection at

all with beetles and Gross can find no signs of
relationship with either *Coleoptera* or *Diptera*.

Embryology and the study of larval forms have
thrown so much light on the ancestry of many
animals, that it was hoped that a microscopic
examination of the larvæ of fleas, in various stages
of development, would produce some facts of importance. In this hope entomologists have, to a
great extent, been disappointed. There seems to
be much similarity between the embryos of beetles,
moths, flies, wasps and fleas. Those who have dwelt
on the likeness of the larval flea to the maggot of a
fly seem to forget that the resemblance to an embryo
beetle is nearly as strong.

The young larva of the flea is very transparent
and the digestive canal, heart and nervous system
are easily recognised. The egg-shell breaker is an
interesting example of the development of a temporary larval structure and it is the only known
instance of such a structure in an insect. There
are no traces of eyes. The antennæ are three-
jointed. They are rather long and slender, being
about one-third as long as the head. The head is
well-developed and the larva has no feet.

The biting mandibles are broad and triangular.
Compared with those of other larvæ they are said
to be more like the mandibles of coleopterous than
of dipterous larvæ. The maxillæ, or second pair of

jaws, are somewhat reduced and rudimentary. The absence of eyes and of legs are points of similarity between the larvæ of fleas and flies. The maggot of a fly has also two pairs of jaws, and a pair of antennæ.

At the tail end of the larval flea's abdomen are two small projections called caudal stylets (Fig. 1). They are strong, recurved, chitinous, structures which prop up the body of the larva when it creeps and wriggles. There are similar props in the larvæ of certain beetles and no exactly similar organs are known in dipterous larvæ. But caudal stylets are of small taxonomic importance.

In one respect the mature flea is certainly nearer to a beetle than to a fly: the three joints of the thorax are free as in a beetle and not fused as in a fly; but when one studies the mouth-parts, the true view seems to be that the mouth-parts of a flea are equally unlike those of a fly and those of a beetle. Such being the present state of our knowledge, one must wait for fresh light to be thrown on the matter by further researches. It seems unlikely that the immediate future will produce a solution of the problem.

## CHAPTER III

### THE MOUTH-PARTS AND SENSE-ORGANS

WHEN the outward anatomy of a flea was described, in an earlier chapter, the mouth-parts, which
form a sort of beak or proboscis under the head,
were mentioned.   These most interesting parts of the
insect must now be dealt with.   The reader probably
knows that some insects have mouths for sucking
fluids and others mouths for biting solids.   A moth
or a fly cannot masticate solids, whilst a beetle or
a cricket has effective biting jaws.

The first naturalist who studied the mouth-parts
of a flea, with such microscopes as were then available, was Leeuwenhoek.   He was a Dutchman who
worked at the end of the seventeenth century, and
the minute accuracy of whose observations still often
fills modern naturalists with wonder.   Microscopic
work was then in its early days, but Leeuwenhoek
clearly made out the two serrated lancets (Fig. 4)
which are called the mandibles.   His " Microscopical
observations on the structure of the proboscis of
a flea" were published in the *Transactions of the
Royal Society* in 1706.

The mouth-parts of fleas are differently constructed
from those of all other insects.   Around the orifice
of the mouth are a number of appendages which

form a complicated apparatus for piercing and sucking. Their construction and use cannot be described without employing some technical terms. When the names of the parts have been mastered, a diagram will make their relative positions clear. It may be necessary, first, to remind the reader who is not an entomologist that the real *mouth* of an insect is the entrance to the alimentary canal, and that the appendages of the mouth, which act like jaws for masticating or like tubes for sucking, are really modified limbs. In fleas the mouth is suctorial. But before sucking up the blood the flea must first pierce the skin of its host. The paired mouth-parts, then, are modified limbs which correspond with those appendages on the thorax of an insect which we call the three pairs of legs.

The primitive insect, of which fleas and all other insects are descendants, was, it is supposed, composed of a succession of segments each bearing a pair of jointed appendages. Insects of the present day never have more than six legs, but the foremost pairs of appendages have been bent round, reduced in size, and altered in shape so as to serve as mouth-parts.

Now the mouth-parts of the flea for which only technical names exist are the maxillæ and maxillary palpi, the labium and labial palpi, the mandibles and the labrum. The labrum is considered by some authorities to be the hypopharynx. It will be best to deal with each of these in turn and then to explain how they act in combination.

*The maxillæ.* These are a pair of horny or chitinous triangular plates one on either side of the flea's face. They are placed some distance away from the orifice of the mouth and to the right and left of it. They do not serve for piercing or sucking, and appear to have no active function unless they serve to separate the hairs of the host and enable the flea to reach the bare skin. In the majority of bat-fleas (*Ceratopsyllidæ*) the maxillæ are dumb-bell-shaped but in all other fleas they are more or less triangular. From the fore part of each springs a palpus. Like other highly chitinised parts of a flea, the maxillæ are usually dark in colour.

*The maxillary palpi.* These are jointed hairy feelers which project forwards and were mistaken by the older naturalists for antennæ. They spring from the base of each of the maxillæ where these latter organs are joined to the head of the flea. The palpi are sense-organs as the number of sensitive hairs on their surface indicates. The maxillary palpi of fleas are always composed of four segments.

*The labium and labial palpi.* These form together what is called the *rostrum* of a flea. The labium is a single organ which projects beneath the aperture of the mouth. It may be described as the lower lip of the flea. At its end it divides into two comparatively long branches. These are the labial palpi. The actual piercing organs, which will be described below, are the mandibles and labrum. They are not so conspicuous as the rostrum which protects them.

When the piercing organs are at rest they are partly retracted. The external portion is encased in the tubular rostrum. The tube is formed by the two labial palpi which are situated at the apex of the short non-divided labium. The number of segments composing each labial palpus in fleas varies, so far as we know, from two to seventeen. In most fleas, however, the labial palpus consists of five segments. This appears to have been the original state of things ·in the ancestral flea ; the palpus with more and the palpus with less segments being derived from the normal five-jointed one. The rostrum of a flea is not a piercing organ like that of a fly and a bug. The two labial palpi separate and lie flat, right and left, on the skin when the true piercing organ is driven into the host. The labial palpi therefore require to be flexible, and this is attained by increasing the number of segments or by reducing the amount of chitinisation or horniness. We shall find in the chigoes and their allies a rostrum which is pale, weak, soft and scarcely horny. Among other fleas where the rostrum is prolonged and strongly chitinised we shall find greater segmentation.

The small bristles at the extreme tip of the rostrum seem to be sensory organs. They are like those at the apex of the maxillary palpus. When a hungry flea is put on one's arm, it appears to test the skin with these bristles before it ventures to make a puncture.

*The mandibles.* These are a pair of sharp lancets with serrated edges. They make the puncture and are interlocked with the labrum to form a sucking tube.

*The labrum.* This is the central portion of the mouth-parts and is in fact a prolongation of the upper lip of the flea. It is a hard, sharp, awl-like instrument : in shape like a horny trough. Its edges are more or less toothed. Its apex is pointed and it is as long as the mandibles.

The general appearance and the relative positions of the mouth-parts are shown in Fig. 4.

Bearing in mind, then, that the piercing organs are the labrum and the two mandibles, and that the rostrum (composed of labium and labial palpi) is merely a sheath, it is easy to form a clear picture of a flea feeding. Anyone who is bold enough to place a hungry flea on the bare skin of the arm can readily observe through a powerful lens what happens. When the flea has chosen a spot to pierce the skin, the rostrum, with the mandibles and long upper lip or labrum inside it, is moved a little forward. The

flea then lifts its abdomen upwards and presses the
piercing organs down into the skin.   In doing this,
it uses its own weight and the strength of the fore-

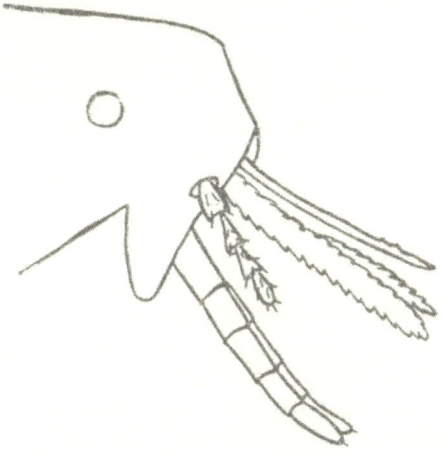

Fig. 4.   Diagram of the mouth-parts of a flea.   The slender
awl-like structure at the top is the *labrum*.   Beneath
are the paired *mandibles* with serrated edges.    The
four-jointed hairy *maxillary palpus* is below, only one
being shown.   Protruding from the base of the face is
the *labium* which supports the jointed *labial palpi*. The
flat obtuse triangular structure from which the palpus
springs is the right-hand *maxilla*.   The left maxilla is
concealed behind.

most and middle pairs of legs.   The hind pair of legs
are lifted up into the air.   The head can soon be
seen coming nearer the skin.   The rostrum then

divides in the middle. The labial palpi are forced apart as the mandibles and labrum penetrate into the victim's flesh. Finally, they are driven entirely asunder and lie flat on the skin of the host, one to the right and the other to the left. The flea then satisfies its hunger. A stream of blood is sucked up, and when the meal is over, there is a forcible action of the legs and the mandibles and upper lip are withdrawn with a jerk. Numerous observers have remarked on the habit possessed by fleas of discharging the contents of their intestines whilst actually engaged in sucking. In many cases a drop of bright red blood is squirted from the rectum during the operation of feeding, and this appears to be a common practice among blood-sucking insects. Its bearing on the feeding operation of the flea has not been discovered. But its possible consequences in transmitting diseases from host to host will be seen in a subsequent chapter on fleas and the transmission of plague.

It is said that the nervous systems and brains of fleas are not so highly developed as those of many other insects such, for instance, as ants, bees and other Hymenoptera. Having drawn attention to the distinction between the external skeleton of a flea and the internal skeleton of a vertebrate, one may with profit do the same in the case of their nervous systems. In both cases the nervous system serves to convey sensations from the sense-organs, and movements to the muscles. In the vertebrate, as the reader doubtless knows, there is a brain, a nervous cord running from it down the backbone, and a number of nerves issuing, from the spinal cord and from the brain, in various directions. Here the main nervous system runs down the *back* of the animal. In a flea, or other insect, the nervous system consists of a chain of ganglia connected by a nervous cord. A ganglion is a nerve centre and, in a sense, each is a brain which may be likened to the one brain of the vertebrate. We have in the cord of ganglia a series of brains, as it were, running from the head down to the extremity of the abdomen. Each ganglion is a mass of nerve cells, from each of which a fibre passes off to unite with the other fibres and make a nerve. The first ganglion in a flea is placed in the upper part of the head above the gullet. It may be called the brain since it receives the nerves of the antennæ and eyes. In the ancestral insect we may suppose that there was a pair of ganglia in each segment. Since the head of the flea consists of several fused segments, we may fairly draw the conclusion that the brain is the result of the fusion of several pairs of ganglia.

The brain of the insect occupies the same position in the body as the brain of the vertebrate; but the rest of the nervous system lies on the floor of the

body *under* the digestive canal of the flea, whereas in
the vertebrate it lies along the back and *above* the
digestive canal.   The dorsal spinal cord of the vertebrate is then a ventral nervous cord in a flea.

The sensory nerves, which transmit sensations
from different sense-organs, and the motor nerves,
which send stimuli to the muscles, take their origin
from other ganglia besides the ganglion above the
gullet.   In bees and some other insects it has been
shown that the nerves from the palpi and mouth-
parts go to the next ganglion which is beneath the
gullet.   The same is probably the case with fleas; so
when we speak of the *brain* of a flea we must remember that it has a relative rather than an absolute
claim to that title.   A flea has really many brains.

In  certain  blind  insects, where the  eyes  are
wanting, parts of the brain are completely atrophied.
Whether this is so in the blind species of fleas does
not seem to have been investigated.

We pass now from the central nervous system to
the sense-organs of the flea.   The chief are the eyes,
the antennæ and the pygidium.   In regard to the
eyes nothing more need be said.   The antennæ are
probably far more important organs to a flea than its
eyes ; but inasmuch as they are at ordinary times
concealed in a groove they are not very conspicuous
(Fig. 5).   The first tolerably accurate plate of a flea
by a naturalist will be found in Hooke's *Micrographia*

(1664). Robert Hooke (1635—1703) was a somewhat
eccentric and irritable man of science who acted as
secretary to the Royal Society. His labours were too
varied to be effective. He nearly discovered the laws
of gravity and also studied fleas. To him belongs
the credit of having detected the antennal groove.
Just as many of the older naturalists thought that

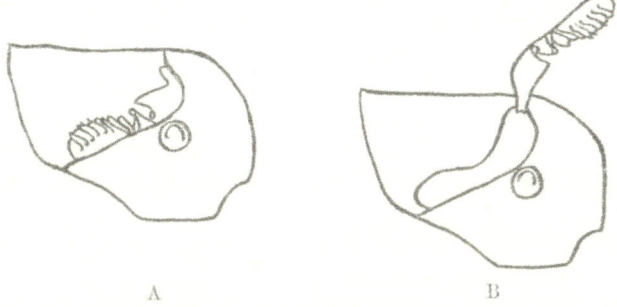

A                                          B

Fig. 5. The antenna of a flea. A, concealed in the groove. B, protruded from the head. The versatile *basal segments* and the
terminal *club*, in this case with segments on one side of it,
should be noticed.

the maxillary palpi were antennæ, so others thought
that the antennæ of a flea were its ears. And when,
with the help of their lenses, they saw the antennæ
erected and protruded from their grooves, they
imagined that the insect was cocking its ears and
listening after the manner of a horse or ass. But
the antennæ of fleas are much more to them than

ears; though it may be that they are also auditory
organs. They are certainly tactile and olfactory
organs as well. In outward structure each antenna
consists of two parts which may be called the stalk
and the club. The club is divided into a number of
segments and is plentifully supplied with hairs. In
some species the cuts which divide the different
segments appear to be confined to one side of the
club. In others a sort of central core holds the
segments of the club together. The antennæ, therefore, are undoubtedly exceedingly complex organs.
Such an insect as a flea may well be far more sensitive
to movements of the air, vibrations of the earth,
smells, light rays and sound-waves than a human
being. In their origin the antennæ, like the paired
mouth-parts, are modified appendages of the fused
segments which compose the head of the insect. The
fact that there are four pairs of appendages on the
insect's head, viz. (1) antennæ, (2) maxillæ, (3) labial
palpi and (4) mandibles has been put forward by
some entomologists as evidence that the head is
formed of four primary segments.

Antennæ apparently enable fleas to find their
bearings, to communicate with one another and to
discover the whereabouts of the opposite sex. But
it is especially as organs of smell that they play
a most important part in the flea's social life. They
enable couples to find one another; and, when the

sexes come together, the antennæ of the male are
usually raised and exposed from the groove.   Insects
generally have some means of cleansing dirt from
their antennæ.   Some make use of their legs, others
of their mouth-parts.   In fleas there is often a row of
short hairs at the hind margin of the groove which
may serve as a kind of comb for cleaning these
delicate organs of sense.   But further observation on
this point would be interesting, for no one appears to
have seen the comb in actual use.   Female fleas are
said usually to carry their antennæ esconced in the
grooves, whilst the males more frequently protrude
theirs.   The antennæ of the males are generally
longer than those of the females.

There are certain noteworthy organs of sense
which appear to exist on the upper surface of a flea's
head and body.   They take the form of small con-
vexities of the body surface, lentil-shaped and each
surrounded at the base by a ring.   Somewhat similar
sense-organs are widely spread through the insect
world.   As to their function, divergent views are
held.   Some think that they are for the perception
of sounds, some for the perception of light rays, some
for the perception of rays of which we are un-
conscious.   Since these organs are placed, at times,
in unprominent parts of the body it seems more
probable that they are affected by sound than by
light.

The preference which fleas show for certain animals, and the repulsion which they manifest on being allowed to suck blood from an unaccustomed host, lead one to believe that they have a sense of taste. This sense in other insects is apparently seated in certain microscopic pits and hairs which form the ends of nerves and are distributed round the mouth. Whether fleas can hear is not, it seems, definitely known.

A large number of fleas possess what is called a frontal tubercle. It is a notch in the centre of the forehead but nearer to the mouth than to the antenna. Sometimes the tubercle projects from a groove. This is most marked in the genus of African fleas *Listropsylla*. The real nature of this organ is unknown. Some regard it as an organ of sense. Its homology is also uncertain. To some it suggests the egg-breaker of the larva and they regard it as a relic of the larval stage. To others it suggests an eye and they regard it as the remnant of an unpaired ocellus possessed by the ancestral flea.

An exceedingly remarkable organ of sense, which is found in all fleas, is called the pygidium. It is a sensory-plate plentifully supplied with hairs and nerves and always placed on the back of the ninth abdominal segment. Of all its uses we are still somewhat uncertain but some observers declare that at the season of love the male flea bestows caresses on the pygidium of the female.

In many species the male flea is sufficiently different in outward appearance from the female to be easily distinguished. The male is usually smaller and the last segments of the abdomen are so shaped as to give the look of a tail tilted into the air. The frontispiece represents a male flea and shows this well. The internal organs of reproduction (testes and ovaries) in the male and female are placed near the end of the abdomen. The seminal outlet and common oviduct open to the rear of the sensory plate on the ninth segment of the abdomen. The external genital armature of the male flea is exceedingly complicated and quite unlike that of any other insect. When the sexes are united, the usual position is reversed, and the male is *beneath* the female.

It is well known to every entomologist that the hinder segments of insects are often modified for reproductive purposes. In male fleas it is the eighth and ninth abdominal segments which are altered. In the females the eighth, and also often a portion of the seventh, has assumed a peculiar shape. The clasping organs of the male flea are portions of the ninth segment and form together a kind of claw reminding one of the pinchers of a lobster. It is used by the male flea in the breeding season to detain and hold the female.

Every entomologist also knows that the external

sexual organs of insects, of both sexes, are of special
importance to the systematist or classifying naturalist.
They often enable him to recognise the species when
other organs do not show sufficiently striking cha-
racters. A minute study of the genitalia of fleas is
an absolute necessity to the systematic entomologist,
the more so as fleas do not present nearly as many,
or nearly as varied, external differences as do the
species of most winged insects where colour and
pattern of wings are both important.

## CHAPTER IV

### THE INTERNAL ORGANS OF A FLEA

A FLEA like every other animal must feed and
breathe, which leads to a consideration of the internal
organs of digestion and respiration. The digestive
canal is a slender tube which connects the mouth
and the anus, and which is less convoluted and much
straighter than in the higher vertebrates. Fig. 6
will show the relative positions of the various parts,
namely, the mouth, pharynx, gullet, gizzard, stomach,
and rectum. Connected with the digestive canal are
certain glands and organs of excretion. The alimentary tube itself passes through the middle of the
flea's body, and is kept in that position partly by

muscles and partly by the numerous branching air-tubes through which the insect breathes. Above it

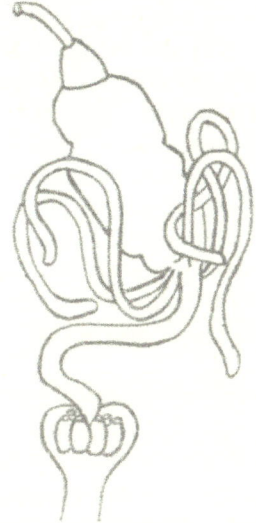

Fig. 6.  Diagram of the alimentary canal of a flea.  At the top is shown the orifice of the *mouth*, leading into the *pharynx*.  Next comes the short *gullet*.  The *gizzard* is the smaller organ immediately before the stomach.  At the base of the *stomach* are four vermiform tubes, which are the *Malpighian tubules*.  From the base of the stomach issues the *intestine*, which leads to the *rectum*, where the six *rectal glands* are shown.

lies the heart, and beneath it the nervous cord or chain of ganglia.

The mouth of a flea, as of any other insect, is merely an orifice which forms the opening into the alimentary canal. Around the orifice are the various mouth-parts which convey blood to the mouth, but these, the reader will doubtless remember, are the modified limbs or appendages of the segments that compose the flea's head. The mouth, then, gives access to the digestive canal. The first part nearest the mouth is the pharynx which merges gradually into the gullet. Here is placed the pharyngeal pump which is provided with a sucking apparatus. Muscles attached to the dorsal part of the so-called aspiratory pharynx cause it to expand and contract, owing to the elastic reaction of its walls. The operating muscles, which do this, are in the head of the flea. When these pharyngeal muscles contract and relax in regular sequence, a rhythmic action of the pharynx itself ensues and a steady stream of blood is forced or drawn from the mouth stomachwards. In a light coloured flea, under a powerful lens, this action may be watched in the living insect.

Behind the pharynx comes the gullet, which leads down to the gizzard. It is perhaps needless to add that this organ, neither in appearance nor in use, bears any resemblance to the gizzard of a bird, which grinds hard food. The food of the adult flea consists solely of liquid blood.

The organ called *gizzard* in the flea, for want of a

better name, is, however, remarkable.  Its function is
not quite certainly known.  It is a bulbous expansion
in the front of the stomach and situated at the junction of the stomach and the gullet.  It contains a
multitude of chitinous finger-like processes tapering
towards their extremities.  From their general
arrangement the complete collection of processes
would act as an effective sort of valve and prevent
the return of the fluids from the stomach.  It seems
most probable that this is their function.  During
the life of the flea the stomach is constantly churning
its contents.  Some valvular arrangement between
the stomach and the pharynx would seem to be
essential; the pharynx is normally collapsed, as the
reader may remember, and its walls are drawn apart
by muscles attached to its exterior.  When the
pharynx is full of blood the muscles relax, the walls
collapse like elastic, and the blood is forced into the
stomach.  In many cases a flea will feed when the
stomach is already tensely full of blood; and some
sort of valve is therefore needed to prevent regurgi-
tation into the pharynx when the pharyngeal muscles
contract and the walls of the pharynx itself are drawn
asunder.

This valvular arrangement at the anterior end of
the flea's stomach has been minutely studied in connection with recent plague investigations, because
there was a theory that fleas carried infection by

vomiting the septicæmic blood from their stomachs and so transferred the plague bacillus to the puncture which they made in the skin.

But an experiment, which has been tried several times, seems to show that the supposed valve is effective. The stomach of a flea which had recently fed was dissected out intact. As long a portion of rectum as possible was left attached at the hinder end. The gullet having been severed, well in front of the valve, pressure was applied with a blunt tool with the object of forcing the blood through the gullet. The hind aperture of the stomach was, at the same time, closed by pinching up the rectum. The result was that, in no instance, was it possible to force blood through the passage which leads into the gullet. Yet sufficient pressure was applied to burst the stomach.

The stomach of a flea is a pear-shaped sack which occupies an appreciable part of the insect's abdomen. That it is capable of containing a comparatively large amount of blood is apparent from the observation that after a flea has enjoyed a good meal nearly the whole of the abdomen is seen to be filled with a bright red mass. During the investigation of the part played by fleas in spreading plague an endeavour was made to measure, as accurately as possible, the average capacity of a rat-flea's stomach when filled with blood. Healthy fleas, taken from Bombay rats, were starved

for twelve hours, and at the end of that time were
fed on healthy animals.   The stomach was then dissected out whole and floated in a salt solution.   Any
adherent organs or muscles were carefully removed.
Under these conditions the stomach can be examined
and measured under the microscope.   The average
capacity of a rat-flea's stomach has been approximately
estimated to be half a cubic millimetre.

The stomach of a flea is therefore, comparatively
speaking, very large.   The blood remains in the
stomach in a partially digested condition.   It gradually diminishes in volume, showing clearly that
absorption is taking place.   At the end of so much
of the digestive process as takes place in the flea's
stomach, the blood has become reduced to a thick,
slimy, dark red mass.   This passes down the intestine
to the rectum, where it is perhaps further influenced
by the secretion of the so-called rectal glands.
Finally, the undigested remains pass from the rectum
in the form of very minute, round, almost black, tarry
drops.

The terminal section of the flea's digestive canal
is called the rectum.   Here are placed the rectal
glands (Fig. 6), which are six in number.   Their
function seems not to be certainly known.

The external opening of the rectum is placed at
the extreme end of the flea's body between the
tergite and sternite of the tenth segment.

We pass now to a couple of quite distinct appendages of the digestive canal, namely the salivary glands and the urinary tubules. In fleas the salivary glands are four in number. Two are placed on each side of the anterior end of the flea's stomach. Each is a simple acinous gland embedded in the body and lined with cells which secrete the saliva. The four ducts from the pairs of glands unite to form two ducts; and the two ducts thus formed run forward and open into the salivary pump. A spiral chitinous membrane lines the inside of the ducts, keeps them distended, and gives them somewhat the appearance of tracheal tubes. The salivary pump is placed quite in the front part of the insect's head, and is an organ worthy of special notice. It receives the saliva from the glands by means of the two salivary ducts which have just been described, and propels it through the exit duct of the pump into the salivary canal in the mandibles. The pump itself is a hollow chitinous organ. Muscles attached to the walls alternately contract and relax, drawing up the salivary secretion and expelling it through the exit-duct. The opening of the exit-duct is adjusted so as to be opposite to the canals which extend down the mandibles like troughs.

It would seem that when the flea is feeding, saliva is pumped into the puncture and blood is pumped out. There is, as it were, an effluent and an affluent stream passing along the mouth parts.

The urinary tubules are excretory organs which
carry off, in solution, the waste products of the flea's
body.    They are sometimes also called Malpighian
tubes (Fig. 6).    This name they received after
Malpighi (1628–94), a famous Italian anatomist, who,
four years after Harvey's death, saw with his own
eyes the capillary circulation of which Harvey had
only inferred the existence.    He also was the
first to detect the urinary tubes of insects.    These
tubules answer to the kidneys of the higher vertebrate.    They vary in number in different insects from
two to over a hundred.    In fleas there are four.    They
are longish, slender, tubular glands which are closed
at one end, but, at the other, open into the rectum.
The urinary excretions come from the blood, pass
down the tubes into the rectum, and so leave the
flea's body by the anus.    In insects the urinary
excretion is, generally, only partially liquid.

The organs of respiration in a flea consist of a
series of tracheæ, or air-tubes, which open by apertures, called stigmata, at the sides of the body.    These
air-tubes branch and form an elaborate system of
ramifications.    They have a horny lining and are
supported by a spirally-wound thread-like thickening.
In this way air is conveyed from the external world,
and the oxygen, which vital processes require, is
conducted to all parts of the insect's body.

The blood-system of a flea is far less complete

than that of the lowest vertebrate. The blood is
almost colourless. A large contractile heart drives
it into the main blood-vessel. There is, however, no
closed system of arteries, capillaries, and veins such
as the higher animals possess; and the blood circu-
lates in the whole cavity which intervenes between
the body-wall and the various internal organs. There
is little need for an elaborate system of blood-vessels
since the internal tissues are supplied with oxygen
by the ramifying air-tubes. Fleas have more of the
air-holes called stigmata than any other insects.
Each of the three segments of the thorax has a pair,
as well as the second to the eighth segments of the
abdomen. The spiracles or apertures lie free on the
outside of the body. In beetles, and other insects
which run through dusty places, they are lodged in
the thin membrane between the segments.

The heart of a flea is a very delicate pulsating
tube which lies along the back, above the digestive
canal and immediately beneath the integument. One
may attribute some of the extraordinary strength
and vital energy of a flea to the fact that, by the
blood-system and the air-system, the tissues of the
body are kept richly supplied with oxygen. The
blood of a flea is a thin fluid and, of course, without
red corpuscles. The blood that is shed when a flea
is crushed comes from the stomach and not from the
blood-vessels of the insect.

The internal organs of fleas cannot be studied without dissection under a microscope. Dissection is best carried on in a solution of salt and water. Fine needles mounted in penholders are the most handy implements. But the point of even the finest commercial needle that can be bought is too blunt for fine dissections, and it is necessary to sharpen it. This can be done by the help of a rapidly revolving emery wheel, varying the inclination of the needle-point to the wheel, so as to grind off the angles. The flea to be dissected is put in a drop of salt solution, on a slide placed on the stage of the dissecting micro-scope. In the left hand should be a needle with a blunt conical point, in the right a needle with an oblique point. The antennary groove of the flea should then be transfixed and held firmly by the left-hand needle.

The point of the right-hand needle is then inserted under the edge of the third or fourth abdominal seg-ments. The segments can then be peeled off by a skilful dissector much as we peel off the skin of a shrimp for our tea at the sea-side. The internal organs of the flea then float off in the salt solution; and by using two very fine pointed needles they can be further separated. It is useful to have one needle ready with a hooked end and another fashioned into a minute knife or scalpel.

The most conspicuous of the internal organs will

be the stomach and intestine. The salivary glands will be found at the side of the stomach with a certain amount of fat round them. Their extraction is not so difficult as might be supposed. The hooked needle can be used to hook the salivary duct.

The most difficult parts to dissect are the organs connected with the mouth and rostrum. It is best to remove the head and transfix it with the left-hand needle, then to scalp the head by removing the dorsal half of the chitinous carapace. A bold plunge with the right-hand needle will sometimes effect what is desired. A pull on the labium will sometimes bring out the pharynx. It must be confessed that successful dissections are often obtained more by good luck than by skilful management. The use of dilute potash solution facilitates the study of chitinous parts by jellifying the muscles.

## CHAPTER V

### THE HUMAN FLEA AND OTHER SPECIES

THE human flea (*Pulex irritans*) appears to occupy an isolated position. The genus *Pulex* which Linnæus established has now been reduced until it contains one species only. The human flea belongs to the group with eyes and without combs. In some respects it is the most specialized of all the *Pulicidæ*. The chigoes (*Sarcopsyllidæ*) resemble it and are doubtless derived from the *Pulicidæ*. The chief structural character of this interesting insect is the greatly reduced thorax. But it can be distinguished from any other known flea by the fact that the upper segment of the hind leg (hind coxa) bears a number of hairs on the inner surface of the posterior portion. A more noteworthy feature in this flea is the presence, in a large proportion of specimens of both sexes, of a small tooth at the edge of the head. This small tooth is sometimes absent; but, when present, both its position and its structure indicate that it corresponds to the fifth tooth in the head comb of the dog-flea (*Ctenocephalus canis*) (Fig. 7). In the hedgehog-flea (*Ct. erinacei*) the teeth of the combs both on the head and on the thorax are small in size and few in number. Occasionally they almost disappear. The conclusion seems justified that the human flea is descended from an ancestral form with combs. To discuss whether the combs became useless and were lost when the host lost the hairy covering of its body would lead into regions of vague speculation and occupy time unprofitably.

The nearest allies of the human flea, which are found on various animals, are all inhabitants of the Old World. The indigenous fleas of America are only distant relatives of *Pulex irritans*. Our know-

ledge of the present and former distribution of this
species is deplorably meagre.  The many books of
travel published in the early part of the nineteenth
century contain hardly any records of fleas.  The
human flea is now cosmopolitan.  Specimens identical
with those from Europe are found almost everywhere.
But it may be doubted whether this was the case
before the great era of travel and steam began in
last century.

There is one strange and, indeed, inexplicable fact
in connection with the distribution of this cosmopolitan species of flea.  It is absent from the oases of
the Sahara and the Haussa countries immediately to
the south of the great desert.  These countries have
long been in communication with places where *Pulex
irritans* is known to abound.  There is no natural
barrier.  The habits of the natives would encourage
fleas to thrive, and other forms of human vermin are
plentiful.  There is, apparently, only one explanation
that is forthcoming.  It is suggested that the soil
and climate in these regions of Africa are, for some
reason, unsuited to fleas.  In other parts of the Dark
Continent, where there are European settlements, the
human flea seems to thrive surprisingly well and
to attack Europeans and natives, as well as wild
and domestic animals.  In those parts of Asia where
there are European colonies and much intercourse
between settlers and Orientals, *Pulex irritans* is a

well-established and thriving parasite. Unfortunately, there is no means of knowing whether this was the case among the native populations before European travellers and traders arrived. *Pulex irritans* has, however, recently been found on the natives of German New Guinea living some 10,000 feet above sea-level and in great isolation. Seaports are everywhere infested with fleas.

Another problem on which no light has been thrown concerns the evolution of the human flea. It would be of great interest to know whether the present species has undergone modifications of form since it became a parasite of the human race; whether we inherited the species from our simian ancestors; or whether the flea of one of the lower mammals became parasitic on mankind. In the Old World this flea is essentially a parasite of man. It occurs only occasionally on other mammals. In America it certainly appears to occur more frequently on mammals, other than man, than it does in the Old World. Human fleas can propagate in deserted human dwellings. The larvæ find nourishment in any refuse that has been left behind, and the adult insect can apparently continue for some time to reproduce itself without a meal of any sort and certainly without human blood. Travellers in the East and in Africa have described how on entering huts in deserted villages they have found their clothing covered with myriads of fleas, sometimes ravenous, and at others weak from long fasting.

The human flea is a good deal more select in the choice of a host than some other species. The cat-flea (*Ctenocephalus felis*) has been found not only on the cat, but also on the dog, tiger, leopard, goat, horse, rat, hedgehog, kangaroo, deer, guinea-pig, rabbit, and on man. Many of these were specimens collected in zoological gardens. Although when hungry and confined in a test-tube the human flea will readily bite a rat or a guinea-pig, it has been found that human fleas kept with no other food-supply than rats and guinea-pigs soon die off.

When large numbers of human fleas were wanted for experiments in Bombay, guinea-pigs were used as traps to attract them. On one occasion two guinea-pigs placed in a house which had been vacant for some days, and in which fleas must have been short of food, failed to attract any of this species; while a man who entered the house shortly afterwards acted as an admirable trap. Those who have not had experience of the abundance and voracity of fleas in oriental countries can hardly believe the numbers of human fleas that may be captured by sending a bare-legged man into a deserted house and then picking the fleas off him. In one house 31 *P. irritans* were taken on a man's legs in a few minutes. In another house 84 *P. irritans*, 8 cat-fleas and

1 bird-flea were caught. In a third, 150 *P. irritans* and 4 cat-fleas were captured in a short time.

The piercing organs of the human flea are strong and well developed. This is rare in a flea which, far from having adopted stationary habits, is a very active insect. It has been suggested, with some show of probability, that the wide and strongly serrated mandibles were acquired after man became the host. The naked skin and rough garment of mankind would render the claws and legs of the flea insufficient to keep the insect in a steady position when feeding. Natural selection would, in due course, strengthen the mouth organs.

The division of mankind into different races, many of which are quite as distinct as the various species of some genus among other animals, leads one to expect various races among the fleas which are parasitic on them. If the sand-martin and the house-martin, the rat and the mouse have distinguishable fleas, one might suppose that the Caucasian and the Hottentot, the Australian native and the Red Indian would follow suit. It may be that further study will show that the human flea now consists of a number of different races. In only one case, however, does a development of this kind in fact appear. Fleas taken off Mexican Indians show slight but fairly constant differences from the true *Pulex irritans*. The specimens are smaller in size, the rostrum is longer and the clasper of the male is more pointed. If the Mexican Indians have a special race of human flea it must have developed after the Indians came to America, or they must have brought it with them when they came. In the latter case this race of flea may still exist in the country whence these Indians originally came.

Apart from this apparently constant race, the individual variation in specimens of the human flea is slight. If a large series of mounted specimens are examined with the microscope, it will be noticed that the bristles or spines on the legs are sometimes more or less numerous. But, with this exception, marked varieties such as are frequently found among other insects seem to be rare.

Although mankind is the true host of this flea, it has been obtained in various parts of the world on various mammals and occasionally on birds. But in England, and probably in other parts of Europe as well, *Pulex irritans* is an undoubted parasite of the badger. A good series of the insect has been got from wild badgers freshly captured near Reading in Berkshire and Hastings in Sussex. In other parts of the world it has been obtained from a variety of small carnivora: cats, dogs, foxes, jackals and polecats. It has also been found on Rodents (*Gerbillus*) and on Insectivora (*Erinaceus*). In South Africa it has been taken off a caracal and in North America off a lynx.

Sandy places such as sea-beaches and picnic grounds, where humanity congregates for pleasure or business, frequently swarm with this species of flea waiting an opportunity to feed. The larvæ are bred in the sand and feed on organic refuse.

The genus most closely allied to that which contains the human flea consists also of a single species only. It is a large flea (*Pariodontis riggenbachi*) found on porcupines all over Africa and in India.

Mankind is, occasionally, bitten by a variety of other species besides *Pulex irritans*. In hot countries the chigoe (*Dermatophilus penetrans*) is a serious and troublesome pest, particularly to bare-footed people. In temperate regions there are rat-fleas, cat-fleas, dog-fleas and bird-fleas which occasionally transfer themselves to man and feast on his blood. But, on the whole, hunger and propinquity rather than free inclination seem to actuate these fleas of which man is only the occasional host. There are besides very numerous species which have never under any circumstances been known to bite man. There is no doubt that some persons are more attractive to fleas than others. The reason for this we do not know. It may depend on the tenderness of their skin, the quality and taste of their blood, or their personal smell, or possibly all three combined.

The various forms of rat-flea which are important in carrying plague from rodents to the human race

are dealt with later on.   Among the commonest fowl-
fleas which bite man are *Ceratophyllus gallinæ* and
*C. gallinulæ*.   Both species infest the nests of many
common passerine birds besides the domestic fowl.
A common parasite of the pigeon is *C. columbæ*,
which also bites man.

Dog-fleas and cat-fleas frequently transfer themselves to man.   It has been asserted that the flea of
the dog and the flea of the cat are indistinguishable.
Several great authorities on fleas, such as Dr Carlo
Tiraboschi in Italy and Mr Carl Baker in the United
States, have maintained that the differences between
*Ctenocephalus canis* and *Ct. felis* were unreliable
and that they are not distinct species.   Mr Charles
Rothschild has, however, shown that the two species
are abundantly distinct.   The *males* of these two
insects can be readily distinguished from each other
by differences exhibited in their respective sexual
organs.   The *females* can be distinguished, at a
glance, by the different shape of their respective
heads.   Fig. 7, which shows the head of a female
dog-flea above and of a female cat-flea below, illus-
trates this.   It will be seen that *Ct. felis* has a much
longer and more pointed head than *Ct. canis*.   In the
*males* the difference in the shape of the head is less
strongly marked, but is quite perceptible.   There are
several minor differences in addition which serve, but
less clearly, to distinguish these two insects.   The

first genal spine, or first tooth in the head-comb, is shorter in the dog-fleas of both sexes than it is in the cat-fleas.   The abdominal stigmata appear to be

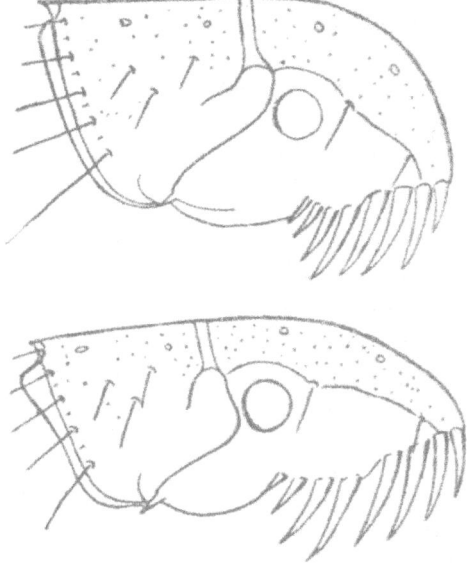

Fig. 7.   The head of a female dog-flea (above) and a female cat-flea (below) to illustrate the difference in shape. In the males the difference is less strongly marked but quite perceptible.   From *Novitates Zoologicæ*, Vol. xii, January, 1905.

larger in a dog-flea than in a cat-flea, and there are differences in the bristles which seem to be constant.

Both species are perceptibly larger than human fleas,
and dog-fleas have always afforded good material for
dissection.  Very few dogs seem to be exempt from
fleas, and the little pets which are carried in ladies'
arms are often swarming with them.

This account of a despised and detested group of
insects would be very imperfect if it did not mention
those educated or performing fleas which have evoked
so much astonishment among people who have
watched them.  It will be best to say, at once, that
the fleas are not educated and that the performance
can only be attributed to their desire to escape.  It
is stated that a performing flea may be broken of the
habit of jumping by being put in a pill-box with glass
sides which is made to revolve like a lottery wheel.  A
short course of this tread-mill teaches the flea that
the objectionable practice of hopping is useless and
exhausting.  It is said that the life of performing
fleas averages eight months, which seems surprising.
They are fed every few days, and the trainers delight
in showing the punctures on their arms where the
swarm of pets has been fed.

Performing fleas are first of all securely fastened,
and this is nine-tenths of the secret, and the art of
education.  A very fine silk fibre is put round the
body and knotted on the back.  The flea may then
be cemented to some moveable or immoveable object.
It may pull a coach by being attached to a pole made

of a bristle. A little paper object stuck on its back is termed by courtesy an equestrian or a ball-dress. The lively imagination of the spectators is of great help. The strength of a flea is wonderful, and on being placed on a sheet of blotting-paper, so that the hooks of the feet get a hold, the coach travels at a fine pace. In the intervals of the performance the coach is turned over, and the performer with its feet in the air does not get exhausted with needless struggles. Or else the fleas are fixed head uppermost, with their legs extended horizontally, to an upright wire driven into the table. Ladies have fans of tissue paper gummed to their limbs. Gentlemen are in the same way supplied with swords made out of fine segments of wire. When two swordsmen are placed opposite each other and the table is knocked they move their limbs. The swords then clash by chance, and we have a representation of a duel not much worse than may be seen in provincial or even London melodrama.

More wonderful are dancing fleas, for there we have a real representation of a ball-room filled with waltzers. The orchestra of fleas, all securely fixed with cement, is placed above a little musical-box. The music proceeds from the box, but the vibrations make the fleas gesticulate violently over their musical instruments. The dancers spin round on the ball-room floor. The couples are fastened by a rigid bar opposite each other, so that they cannot touch or part. Each is pointed in an opposite direction, and tries to run away. A rotary motion ensues which, to the spectators if not to the fleas, is very like waltzing.


# CHAPTER VI

## THE CHIGOES AND THEIR ALLIES

THE chigoes and their allies belong to a group of fleas sufficiently remarkable to deserve a somewhat detailed account. The reader may remember that they form a family to which the name of *Sarcopsyllidæ* has been given. They are the most completely parasitic of any fleas; and the South American chigoe (*Dermatophilus penetrans*) enjoys the distinction of being the first foreign flea ever described. This pestilent insect, of which the female has the habit of burrowing into the flesh of the host, soon made itself known to the early travellers in the tropics of America. Oviedo, the Spaniard and historiographer of South America, in his *Historia General y Natural de las Indias* (1551), seems to have been the first European author who mentions it. After this the chigoe is referred to by writers of various nationalities in many works which were published during the sixteenth, seventeenth and eighteenth centuries. It is an insect which appears under a vast number of different names: chigoe, chigue, chego, chigger, chique, jigger, pico, sico, migua, nigua, ton, and tschike are synonymous.    Catesby in his *Natural History of Carolina* (1743) gives a figure of the insect, which is easily recognisable.    Linnæus, in 1758, described the chigoe as *Pulex penetrans*, and apparently did not know much of its appearance beyond what he learnt from Catesby's picture.    This species and the human flea were the only two which the great Swedish naturalist distinguished by a name; though, under the title *Pulex irritans*, he includes a number of different species such as the fleas from the dog, cat, rabbit and fowl.    The chigoe remained the only member of the family known to scientific entomologists until the year 1860.    An allied insect was then found on a South American parrot.    A third member of the family was soon after discovered, and is note-worthy because it was the first species recorded from the Old World.    It is now known to infest the domestic fowl in all warm countries where these birds have been introduced by man.    A fourth species was collected from a South American bat.    Up to the present time some fourteen different species (belong-ing to three very distinct genera) have been described, and there cannot be the slightest doubt that, when collectors in hot countries turn their attention to the matter, a great many other forms of this interesting family of fleas will be found.

The chigoes and their allies are of special interest for more than one reason.   The females are to a greater or less degree stationary ; they fix themselves firmly to their hosts and become veritable parasites. Several of the earlier zoologists believed that the animal was a mite; and it is somewhat remarkable that Oviedo himself should have so promptly detected the relationship of the insect he saw with the fleas. By reason of the parasitic habits of the females, more is known about their appearance and life than in the case of the more active males.   In some species the males remain, for the present, quite unknown: and not very much is recorded about the early life-history, eggs, larvæ and pupæ of either sex.   The parasitic habits of the chigoes and other allied fleas lead one to expect peculiar modifications of form such as are usually to be observed when an animal passes from an active to a stationary life.   These modifications are the more easily understood as the various species are not all stationary to the same degree.   It is fairly plain that this family of fleas is a development from the less specialised and less parasitic family *Pulicidæ*. In fact the gradual development of the organs from a generalised to a more specialised stage is strikingly shown in these insects.   To follow this in detail would, however, require a very minute and technical know-ledge of their form.

The chigoe family is so well characterised that

a student of fleas cannot possibly have any doubt
whether a flea belongs to this family or not.   Yet
there is great diversity in general appearance, as
well as in details of structure.   One very peculiar
character, namely the enormous swelling of the
abdomen in pregnant females is, moreover, shared
with certain other fleas.   The most distinguishing
character of the family, however, is the rostrum.
This organ, which it may be well to remind the
reader, consists of the under-lip and the labial palpi,
sheaths the piercing and sucking mouth-parts.   In
the chigoes and their allies the rostrum is reduced,
not in length, but in stiffness and in number of
segments.   In this group there are never more than
three segments to the rostrum, whilst in the main
group of fleas, with one or two exceptions such as
the rabbit-flea (*Spilopsyllus cuniculi*), there are
never less than five.   There is no indication of a
comb on the head, but all the family, without exception, have a large triangular projection, which is
more or less curved backwards, at each side of the
head.   These two organs doubtless discharge the
same functions as the combs of other fleas, and
prevent the insect from slipping back as it works its
way through the fur or feathers of the host.

The thorax of a chigoe is exceedingly short.   Two
reasons for this may be suggested.   In the first place,
the jumping power of these fleas is very small and

the muscles in the thorax are consequently reduced.
In the second place, the value to the insect of a
contracted thorax is obvious; for the abdomen of a
chigoe fixed on or in the skin of an animal does not
project so much as it would were the thorax of normal
length.   The troublesome parasite is, therefore, less
likely to be rubbed off by the host.

In most fleas the piercing organs of the mouth
(upper lip and mandibles) are directed obliquely
downwards.     In the chigoes they are directed
obliquely forwards.    It has been suggested that
this forward movement of the mouth is connected
with the stationary life which the females assume.
Fleas which fasten themselves permanently to the
skin of their host, do so in a manner similar to ticks.
The mouth-parts are in a line with the longitudinal
axis of the body.   This attitude, so far as we know,
is assumed by the females of all the family.   The
females of one genus, *Dermatophilus*, actually go head
foremost right into the skin of their host.   The shape
of the head is also beautifully adapted to enable the
insect to fix itself firmly in a tick-like posture.   The
fore-part is remarkably obtuse, and almost has the
appearance of being truncate and abruptly cut off.
When the piercing organs have been thrown forward
horizontally it must be a great support to the insect,
which is fixed by them, if it can press its head down
firmly against the skin of the host.   The wider the

extent of forehead which can be pressed against the
skin of the host the less the strain on the upper lip
and mandibles, which serve as anchors, when the host
scratches.

All the chigoe family have eyes; but in one
recently discovered species the eye is very small
and devoid of pigment. Like those of other fleas,
the antennæ fit into grooves at the side of the head,
and the club, which is the sensitive part of the organ,
consists of eight segments. In a good many fleas the
antennæ are different in the two sexes, but there is
no obvious sexual distinction in this family.

The peculiar development of the mouth-parts is
one of the most singular features in the structure
of the chigoes and their relatives. These important
organs are modified in a fashion not to be found in
any fleas outside the family. Here, as in other fleas,
the mandibles are piercing organs which penetrate
the skin of the host, the upper lip serving in con-
junction as a sucking tube. In ordinary fleas these
organs are retracted when the insect has done its
meal; in the present family they remain, in the case
of the females, apparently permanently fastened in
the skin. The piercing organs are broader and the
serrate edges of the mandibles more solid and heavy
in this family than in the case of other fleas.

The two methods by which fleas keep in touch
with their hosts have already been alluded to. The

two main fixing and clinging organs are the mouth
and the claws. Weak mouth-parts accompany strong
legs. We observe, accordingly, two lines of development. The chigoes and their allies present an extreme
case: for the legs are practically useless for holding
on. The bristles, and the claws as well, are exceedingly thin. In this family the mandibles serve the
purpose of claws. The other line of development is
best seen in a genus of fleas from South America
(*Malacopsylla*), where the piercing organs are short
and weak, whilst the claws and bristles of the legs
are very strongly developed.

The modification found in the rostrum of the
chigoes has already been referred to, and the explanation of this will now be understood. The
rostrum is a sheath, on either side of the piercing
organs, consisting of an under lip and two labial
palpi. When the flea sucks, the labial palpi are
pushed asunder, as the piercing organs are driven
in, and lie flat on the skin of the host. In this family
the rostrum is almost white in colour and soft instead
of being horny or chitinised. Where the rostrum is
strongly chitinised or very horny the flea has to use
a certain amount of force to counteract the spring-
like action of the labial palpi. It is conceivable that
rigid labial palpi would prove inconvenient to fleas
which remain permanently attached to their host by
their mouth organs.

The swollen abdomen of the female chigoe is a strange and a conspicuous object, which is not, however, found occurring to the same extent in all the members of the family. Having burrowed into the flesh of a man, or other mammal, the pregnant female swells and enlarges until she reaches the gigantic

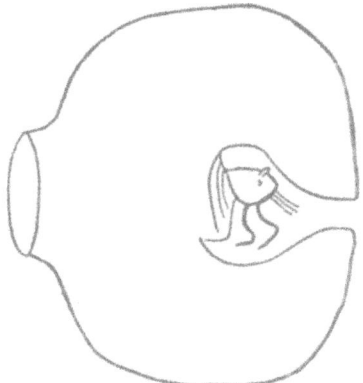

Fig. 8.   Pregnant female of *Dermatophilus cæcata*, a South American chigoe which burrows into the flesh of the host.   The abdomen swells until it surrounds the head and thorax, which are shown in the centre.   The natural size is about equal to a small pea.

proportions of a small pea. Itching and inflammation ensue unless the whole insect be skilfully removed with a needle. To such an extent does the abdomen swell that the segments and the horny plates are driven asunder and the connecting membrane

between is exposed to view.  In the extreme case
of *Dermatophilus cœcata* from South America the
abdomen swells until it completely envelopes the
head and thorax after the manner shown in Fig. 8.
Dr Enderlein found seventeen specimens of this
species in the skin behind the ears of a rat from
Brazil.

The belief that the eggs are laid in the flesh of
the victim is mistaken.  The hind segments of the
body and their stigmata are always exposed to the
air.  The stigma of the eighth abdominal segment is
particularly large.  As soon as the eggs have been
laid, the body of the mother dies, withers, and falls
away from the skin of the host.  The fact that several
females are often found where one has buried herself,
led to the notion that these parasites bred in the
wound.  The truth seems to be that other chigoes
are attracted to a spot where inflammation has made
it easy to burrow.

Chigoes love warmth and drought.  The deserted
huts of natives swarm with them if they are dry.  It
is always said that newcomers are more liable to
attack than natives; but the explanation of this
seems to be that they do not understand the significance of the slight pricks which are felt when the
chigoe fixes itself.  Once the parasite has got under
the skin no pain is felt unless the wound is inflamed
by scratching.  The tender flesh under toe-nails is a

favourite spot of attack. The only remedy is a sharp knife and a little antiseptic wash. Pigs and fowls are sometimes killed by chigoes, and Indians occasionally are attacked by lockjaw after the parasite has been removed. But this is not directly attributable to the chigoe. The eggs are laid one by one; when this operation is completed the mandibles weaken and the shrivelled body of the insect can be rubbed off. But a painful sore may be produced if the parasite is forcibly broken off and the mouth-parts are left in the wound.

The chigoe (*Dermatophilus penetrans*) is a native originally of South America. It ranges from Mexico to Northern Argentina. Some time after the middle of the nineteenth century it was, somehow, carried across the Atlantic and introduced into West Africa. From there it has now spread across the Dark Continent to the Great Lakes, and has even reached Madagascar. Such are the modern facilities of transport which parasites are quick to take advantage of.

## CHAPTER VII

### FLEAS AND PLAGUE

In order to understand the part played by fleas in the transmission of plague it is necessary to have some clear elementary knowledge of the nature of

6—2

that disease. Plague is an infectious fever caused by a specific bacterial organism. *Bacillus pestis* was first identified in 1894 by Kitasato, a Japanese, and immediately afterwards, but independently, by Yersin. It is an exceedingly minute, short, moderately thick, oval bacillus, with rounded ends. It has the most astounding power of rapid multiplication. Nothing is, at present, known of its natural history outside the body of the sufferer, but it can be cultivated. Little is known of its toxic action, but a weak toxin has been got from cultures. The bacillus itself is not of a resistant nature and is easily killed by heat and ordinary germicides. Acids appear to be fatal to it.

In ordinary cases the bacillus is found in buboes. A *bubo* is nothing more than an inflamed gland. In so-called septicæmic cases it is found in the blood of the animal afflicted by the disease. In pneumonic cases the bacillus may be found in the sputum of the patient. It is the custom to speak of (*a*) bubonic plague, (*b*) septicæmic plague, (*c*) pneumonic plague, as though they were three diseases. This is inaccurate: for they are only forms, with varying symptoms, of one and the same disease caused by the same bacillus.

The disease which we call plague is, in truth, really a fight between the afflicted animal and the invading bacillus. It may be inferred from the fact that bacilli are scarcely ever found in the blood in bubonic cases that the invaders are stopped by the lymphatic glands next above the point of inoculation. In such cases the fight, which is the illness, takes place chiefly in the bubo. In non-bubonic cases the fight goes on in the blood-vessels or in the lungs as the case may be.

Whether the plague is primarily a disease of rats would be difficult to say ; but rats and other rodents are very susceptible to it. It has also been transferred to mice, rabbits, guinea-pigs, squirrels, pigs, sheep, goats, cattle and horses. Men and monkeys are equally susceptible. Cats and dogs have been known to die of it and during the Great Plague of London many were destroyed under the belief that they were bearers of infection.

That plague among human beings was associated with mortality among rats and mice, is an observation of great antiquity. The student of the Hebrew scriptures will remember the Book of I Samuel vi. 4 : "Then said they, What shall be the trespass offering which we shall return to him ? They answered, Five golden emerods [buboes] and five golden mice [rats] according to the number of the lords of the Philistines : for one plague was on you all and on your lords."

Eastern authors, of a later date, refer in several places to rats, in times of plague, staggering about as though they were drunk. The Mogul Emperor

Jehangir in his diary of the plague at Agra (1618) mentions the unusual mortality of the rats.  In India it seems long to have been a custom, dictated by experience and caution, to leave houses when rats began to die.  In Europe, during the middle ages, the mortality of rats when the plague was raging does not seem to have impressed the chroniclers and during the recent outbreak at Glasgow (1900) none was detected.

As an illness of mankind, the plague reached Europe from the East.  We have no evidence of any outbreak in Europe before the reign of the Emperor Justinian.  When it raged for the first time at Constantinople (A.D. 542) the mortality was enormous. Ten thousand persons are said to have died in a day with all the symptoms of bubonic plague.

It spread swiftly through the Roman Empire.  In the fourteenth century the same disease under the name of the Black Death again ravaged Europe. Again the mortality was enormous.  Millions perished little suspecting that fleas could be connected with their fate.  Everywhere popular tradition reported the plague as the most highly contagious of all diseases.

In the history of science the plague epidemics in Egypt between 1833 and 1845 are of importance, because the disease was, for the first time, seriously studied by skilful French physicians.  Some of the

French medical school went so far as to deny contagion altogether. The modern view is that aerial infection may be put aside as almost impossible except in pneumonic cases ; but that plague may be transmitted by any method which inoculates the blood with *Bacillus pestis*.

Our modern knowledge dates from the year 1894 when the plague reached Hong Kong. Its existence as a rat disease was recognised. In the autumn of 1896, when plague broke out in India, the men of science, who made careful observations on the spot, were struck by the fact that infection spread from house to house in a fashion that seemed inexplicable, unless the bacillus was carried by an animal.

We pass now from rats to fleas. That fleas might be connected with the spreading of plague was suggested in the year 1897 when Ogata first found bacilli in fleas. He obtained fleas from plague-sick rats. These he crushed, and injected the liquid into a couple of mice. One of these died of plague in three days. The German Plague Commission in Bombay found plague bacilli in fleas, but, for various reasons, did not consider that the bite of the flea was the means by which the disease was transmitted.

The real credit is due to Simond, a Frenchman, who worked during the Indian epidemics. He took fleas from infected animals and observed in their

stomachs bacilli identical with *B. pestis*. He suggested
that the bacillus was carried from rats to men ; and
he brought forward some evidence tending to show
that infected fleas could transmit infection by biting.
But Simond was not able to bring forward conclusive
proof. He pointed out a line of research to others
which has proved exceedingly fruitful. In the same
year (1898) Hankin suggested that some biting insect
might be the means of transmission from rats to man.
The bacillus of plague has now been identified in
ants, bugs, and flies as well as fleas. It seems likely
that any suctorial insect which feeds on a plague-
stricken rat will take numbers of the bacilli into its
stomach.

The points which Simond wished to establish were
that plague-stricken rats with fleas are exceedingly
infective, that they cease to be infective when they
have been deserted by their fleas, and that fleas which
infest rats will transfer themselves to man. Since
1905 an elaborate series of observations and experiments have been carried out. Post-mortems have
been made of countless rats. Numberless fleas have
been collected and dissected. But this summary
would be very incomplete if it did not mention the
work of Verjbitski, a Russian doctor at Cronstadt,
whose labours remained almost unnoticed although
he made his experiments as long ago as 1902–1903.
His thesis, written in Russian, was not published in

any scientific journal. But his ingenious and careful experiments showed that fleas could transmit plague from animal to animal. He found that the commonest flea captured off rats at Cronstadt was *Leptopsylla musculi*, the usual host of which in other places is the mouse. Now this flea does not, except very rarely, bite human beings, and the real significance of the facts discovered was not appreciated.

The common rat-flea in most parts of Europe is *Ceratophyllus fasciatus* and in India and sub-tropical countries *Xenopsylla cheopis*. This last species has acquired the title of "the plague flea," or, more accurately, the oriental rat-flea.

During the plague investigations in India many careful experiments have been made proving beyond doubt that the disease may be transferred from rat to rat by the transference of fleas from a septicæmic to a healthy animal. It was first shown that when fleas were present the plague could be transferred from rat to rat, kept in proximity, but carefully screened so as to avoid any possibility of contact. Next, fleas were collected from rats dead or dying of septicæmic plague and transferred to healthy rats living in flea-proof cages. More than half of the healthy rats contracted plague. It was shown that if fleas are present, the disease once started spreads from animal to animal; and it would seem that the rate of progress was in direct proportion to the number of fleas present.

The blood of a plague-infected rat may contain an enormous number of plague bacilli. Although such figures do not convey any very clear idea of numbers, as many as a hundred million bacilli have been found in a cubic centimetre of rat's blood. A rat-flea, with a stomach of average size, might receive therefore as many as 5000 germs into its stomach ; and it is clear that fleas feeding on a large proportion of plague-infected rats just before death would be almost certain to imbibe at least some plague bacilli. There is, moreover, good evidence for believing that multiplication of the plague bacilli may take place in the flea's stomach. Nor does the blood imbibed by the flea cease to be infective when it passes from the stomach. Both the contents of the rectum and the excrements of fleas taken from plague rats often contain abundant and actively virulent plague bacilli. A number of infected fleas are put into a test-tube : the mouth of the tube is covered with a glass slide, and the mouth is turned upside down. The fleas are then seen to run over the slide, and, in a short time, they deposit an appreciable amount of fæcal matter on the surface. This under the microscope is seen to be covered with plague bacilli; and a large percentage of guinea-pigs, who have an emulsion of the fæcal matter injected into them, contract plague.

It is remarkable that, so far as we know at present, the plague bacillus is confined to the flea's alimentary canal. On rare occasions it is found in the gullet when fleas have been killed immediately after feeding on septicæmic blood. But no plague bacillus has been found in the body-cavity or in the salivary glands.

In the stomach of the flea, plague bacilli have been found in vast numbers twelve and even twenty days after the insect has imbibed septicæmic blood. It is naturally of great practical importance to know how long fleas taken from plague-infected rats remain infective: that is to say, are capable of transmitting the infection to healthy animals. Two series of careful experiments, made during the epidemic plague season in India, have shown that fleas could remain infective for as long as fifteen days. In a third series of experiments, made during the non-epidemic season, it was found that the fleas remained infective for only seven days.

It has been ascertained that both the male and the female oriental rat (*X. cheopis*) flea can transmit plague.

We come now to one of the most interesting questions of all: namely, the method by which the rat-flea transmits plague to a healthy animal.

A variety of suggestions have been made, several of which can be shortly dismissed. It was thought, at one time, that infection might be conveyed by the animal eating the infected fleas. But it is very

improbable that this means of infection is of any
real importance, even if it may sometimes occur.
Experiments in feeding have shown that an animal is
unlikely to become infected by swallowing material
containing plague bacilli, unless the amount is considerable.  Moreover we know that infected fleas
confined in test-tubes readily convey the disease when
allowed to bite an animal.   In such cases the situation
of the primary bubo corresponds with the area of
skin upon which the fleas are placed.  That the
transmission of plague is due to the *bite* of the flea
seems abundantly clear.

It has also been suggested that the proboscis of
the flea acts as a mechanical instrument for the
transference of the bacilli.  No doubt the outside
surface of the flea's proboscis must become con-
taminated, when it sucks the blood of a plague-
stricken rat ; but it is difficult to suppose that
contamination of the proboscis can explain cases of
continued infectivity during which the flea has been
feeding regularly upon healthy animals.

Next, there is a hypothesis that the salivary
glands of the flea become infected and that the bacilli
are inoculated along with the saliva.  The reader
will remember that when a flea sucks, a stream of
saliva is pumped down the mandibles into the puncture.
But this hypothesis is shattered by the fact that plague
bacilli are apparently confined to the alimentary

canal of the flea, and that they have never been
found in the salivary glands.

An apparently more probable explanation, that
the contents of the stomach (in which as we know
the bacilli may multiply) are regurgitated and transferred to the wound by the mouth-parts, is rendered
less credible when we remember that there is a
valvular arrangement at the opening of the flea's
stomach which seems to make such a thing impossible.

Lastly, there remains the only theory on which
we have positive evidence.  It is the theory that the
bacilli contained in the fæces of the flea are deposited
on the skin and then find their way into the wound
made by the piercing organ.  They may be helped in
this by the rubbing and scratching which follow on
the bite of the flea.  We know, of course, that plague
bacilli are present in abundance in the fæces of fleas
taken from plague-sick rats, and that such fæces
are infective to guinea-pigs both by cutaneous and by
subcutaneous inoculation.  Experiments were made
to discover whether the pricks made by fleas were of
sufficient size to allow plague bacilli to enter the
body, no other damage to the skin being done.  Healthy
fleas, confined in a test-tube, were allowed to feed on
a small part of a guinea-pig's abdomen, the hair of
which had been cropped close without injuring the
skin.  Immediately afterwards a few drops of the
septicæmic blood of a rat which had died of plague,

or of a virulent culture of plague bacillus, were lightly spread over the part. Many successful infections were obtained in this way.

Similar experiments were made in which the plague culture was first spread on the skin, and, afterwards, healthy fleas were allowed to feed on the same spot. Successful infections were also obtained by this means.

Two facts then seem to be demonstrated beyond doubt : first, that the puncture made by a flea will allow the bacillus to gain access to an animal's body and to infect it ; secondly, that there is a possibility of infection by the fæces of fleas.

As to whether this is the usual process the highest authorities are not ready to express any opinion. The safest course appears to be to kill fleas but to avoid rubbing them in.

Good work was done during the recent outbreak of plague in San Francisco when the energies of an army of men were directed to controlling and destroying the rat population. Enormous numbers of rats were killed, their breeding places were destroyed and everything was made as uncomfortable for them as possible. Men of science were at the same time engaged in collecting and examining the fleas from many thousands of rats. The great success of the work confirmed the soundness of the theory on which it was based. The spread of the most terrible of epidemic diseases was controlled and prevented by knowledge. At San Francisco the fleas of man, rats, mice, dogs, cats, ground-squirrels and gophers were studied. It was found there, as elsewhere, that while each species of flea has its particular host few are unwilling or unable to attack man and other animals when the host dies.

There is good reason to believe that during the last outbreak of plague in Manchuria the fleas carried the bacillus from the marmots (*Arctomys*) to man.

Plague can be transmitted by the human flea ; but it may be doubted whether this often occurs under natural conditions. The rat-fleas seem inclined to take more readily to mankind than the human fleas do to rats. Experiments at Bombay seemed to show that, though the human flea was able to transmit the plague infection, it does not transmit it as readily as the oriental rat-flea. An explanation of this was obtained when it was discovered that *Pulex irritans* does not live well either on rats, or on guinea-pigs, which were the subjects of the experiments. A count of the fleas was made, each day, in a number of experimental cages, in which live human fleas were placed in company with wild Bombay rats. A great number of human fleas were put into a flea-proof cage along with a rat. Each day a census was taken of the fleas still alive. After twenty-fours hours it was found that little more than one per cent. of the fleas put in could be recovered, and no fleas were ever found alive after the fifth day.

The European rat-flea (*Ceratophyllus fasciatus*) seems to be quite as readily able to transmit plague as the oriental insect. How far other fleas are able to transfer infection we have little or no knowledge. But twenty-seven experiments to transmit plague from animal to animal by means of cat-fleas (*Cteno-cephalus felis*) were once made and none of these were successful. The reason for the failure we do not know.

If infected fleas are kept in captivity after they have fed on septicæmic blood, it is found that, after a while, they are no longer able to convey infection. On being dissected no bacilli are found in them. A clearing process, therefore, evidently goes on. If a number of fleas be fed on a septicæmic rat and, subsequently, be kept under observation and nourished on healthy animals, the proportion found to be infected steadily diminishes day by day. It is remarkable that the existence of numerous plague bacilli in the stomach of a flea does not seem materially to affect the insect's life. Fleas, in other words, do not suffer from plague though they can transmit it.

## CHAPTER VIII

### RAT-FLEAS AND BAT-FLEAS

THE chief conclusions arrived at, as the result of the investigations, during the years 1905 to 1909, into the mode of spread of plague in India, may be briefly stated in the following fashion: The Advisory Committee, under whose direction the investigation was carried out, consider that: firstly, in nature, plague is spread among rats by the agency of rat-fleas ; secondly, bubonic plague is not directly infectious from man to man ; thirdly, in the great majority of cases, during an epidemic of plague, man contracts the disease from plague-infected rats through the agency of plague-infected rat-fleas ; fourthly, where there are annual epidemics they occur during some part of that season when the prevalence of fleas is greatest.

That being so, it is manifest that an accurate knowledge of rat-fleas, their forms, their habits, and their life-history may prove of great importance.

Three species of the genus *Mus* follow quickly in the wake of civilized man and establish themselves all over the globe. They may all be looked upon as more or less domestic animals. The house-mouse (*M. musculus*) is familiar everywhere. The

old black rat (*M. rattus*) chiefly infests ships and
seaports. The brown rat (*M. norwegicus*) is the
most aggressive and distinctive. But all three, by
accidental transference from port to port in ocean-
going vessels, have become distributed over the
world. Their fleas, to a limited extent, have become distributed with them. In connection with
the spread of plague these three small rodents are
of prime importance ; and not less important are
the fleas which are parasitic on them.

In California, the ground-squirrel (*Otospermo-
philus beecheyi*) has been proved to play an important
part in plague infection ; and a full account of its
fleas, and of experiments in transferring rat-fleas
to squirrels and squirrel-fleas to rats, has been
published by American naturalists.

In 1903 Dr Blue, who was in charge of measures
for suppressing plague at San Francisco, observed
that an epidemic disease was killing the ground-
squirrels in the country round San Francisco Bay.
It was shown, somewhat later, that the mortality
among the squirrels was caused by plague, and there
can be little doubt that it was transferred from rats
to squirrels. In harvest time rats migrate to the
fields and use the same runs and holes as the
squirrels. Under these conditions a transfer of fleas
from rats to ground-squirrels is almost certain to
ensue. Two species of flea have been recorded

from the Californian ground-squirrel, and both are parasites of rats. The chain of evidence is really complete, for those who have made a business of hunting ground-squirrels testify to the readiness with which fleas will leave a dead squirrel and bite a human being. In the records of plague in California there are several cases in which there seems to be very little doubt that the disease resulted from handling plague-infected squirrels.

Fleas being wingless insects travel with considerable difficulty over the ground ; and though their hopping powers are notorious they are unable to make any long-continued progress in this way. The methods by which they get dispersed are of interest. Some may be carried by the host in its natural wanderings. Rats appear to be constantly picking up and dropping fleas. Sick rats harbour more fleas than others and therefore more frequently drop them. A hundred fleas have been collected off one plague-sick rat ; and, as we know, if this rat was moribund, some of these fleas would most likely be infected. It is obvious that a plague-sick rat may travel about leaving as it wanders a trail of infected fleas behind it. Rats, too, are frequently transported with certain kinds of merchandise and carry their fleas to the most distant parts of the globe, travelling with all the speed and luxury which modern steamships afford. Rats will dive into sacks of grain or

bran and hide, so that the bag can be loaded as
cargo without anyone suspecting the presence of a
rat inside. *M. rattus* and its fleas, from the habits
of the host, are especially likely to be transported
in this way. Besides, many fleas are now dispersed
without their hosts in merchandise of various kinds.
They may travel great distances in these days of
rapid transport, though adult fleas, without a host
to feed on, generally die in about five days. But
larvæ, which eat organic rubbish, and pupæ, which
do not eat at all, might arrive alive at the end of
a journey of well over a month. On arrival, they
would seek their true host, or the next best available
animal. Not having yet fed, and being newly emerged,
they might survive as long as a fortnight without a
suitable host.

Fleas dislike damp breeding places, but dirty
carpets, chopped straw, old sacking, paper shavings,
and such-like rubbish suit them admirably.

*M. rattus* is fond of making nests on grain bags
and in such sacking the larvæ of fleas are often
found. Where trade is carried on in sacks and
gunny bags this means of distributing fleas and
plague should be kept in mind.

Some rat-fleas, as we know, will feed on man
as well as on rats; but their behaviour is rather
different when they feed on rats and on man. It
has been repeatedly noticed that the fleas were

much more readily attracted by the rat than by man. Although the fleas jump on to a man's hand they take some time to begin to feed. They crawl about and seem to have some doubt where best to begin their sucking operations. Also it has been observed that the fleas much more readily fall off a man's arm, when he moves, than they do when a rat moves. It seems that they are able to get a firmer hold on the rat than on a man ; and it is of interest to note in this connection the larger claws of the human flea compared with the claws of the rat-flea.

It has been shown, in various parts of India, that the number of rat-fleas found on rats varies with the seasons. This seasonal variation of rat-fleas corresponds in a general way with the plague mortality. During the season when plague is bad the average number of fleas per rat is above the mean. During the non-epidemic season it is below the mean. The height of the epidemic corresponds fairly closely with the season of maximum flea prevalence.

Nineteen species of *Pulicidæ* are more or less habitually obtained by collectors on rats and mice. But the great majority of these may be called casual visitors. Six species of *Sarcopsyllidæ* are also occasionally found on rats. These are the burrowing chigoes and their allies which usually attack the head and ears of rats.

The species of flea commonly found on rats are
five in number, and the readiness with which they
bite human beings has been carefully studied.

1.  *Xenopsylla cheopis*.  This is the oriental rat-
flea first described by Mr Charles Rothschild from
specimens collected in Egypt.  The true home of
this flea appears to be the Nile valley, where it may
be found in plenty on various hosts.  Many of these
are desert animals and the flea shows a preference
for rodents.  Having been distributed all over the
world by rats, it now occurs, occasionally, in all warm
climates.  It is the common rat-flea of the tropical
and sub-tropical world.  In India it often happens
that the whole of the fleas collected from rats prove
to be of this species.  But it cannot, apparently,
flourish in cold countries.  In the warmer temperate
zones, such as the Mediterranean and Australian seaports, it occurs in varying proportions according to
the time of year.  The numbers decline with cold
weather.  It readily bites man and is more active
than any other flea in the transmission of plague.
For this reason it is sometimes spoken of as "the
Plague-flea."  It is a smaller and a lighter coloured
insect than the human flea.

2.  *Ceratophyllus fasciatus*.  This is the common
European rat-flea.  It is the rat-flea of the temperate
as opposed to the hot countries of the world.  It
is commonly found on black and brown rats in the

British Islands and the other countries of Northern and Central Europe. It readily bites man, and there is no reason to suppose that, other conditions being equal, it would not be as efficient an agent in spreading plague as the last species has been shown to be in India.

3. *Ceratophyllus anisus.* This is a closely allied species of rat-flea which replaces the last in China and Japan.

4. *Leptopsylla musculi.* This is the mouse-flea and it is as widely distributed over the globe as its host. From mice it frequently moves to rats, and it has been found on them in various parts of Europe, America, Australia, and Japan. It occasionally bites man, but evinces little inclination to do so.

5. *Ctenophthalmus agyrtes.* This flea is commonly found as a parasite of voles and field-mice. When farm-rats take to an open life in the fields they pick up this species from the rustic rodents. In Hertfordshire, Hampshire and Suffolk one half the fleas from rats, collected in farmyards and hedgerows, were found to belong to this species; but whether it is as common on rats all over England is unknown. It appears not to bite man. A closely allied flea (*Ct. assimilis*) is found in central Europe on field-mice and equally on rats which live under the same conditions. It has not been found in England.

The principal occasional parasites of rats are

dog-fleas, cat-fleas, fowl-fleas, and human fleas. The proportions in which they and rat-fleas are found vary greatly in different parts of the world. For instance, in San Francisco nine per cent. of the fleas collected from rats have sometimes been found to be human fleas ; whilst in Italy as many as twenty-five per cent. have been identified as cat- and dog-fleas.

It must be borne in mind that when new countries are opened up by man the rats, which follow in his rear, exterminate numbers of the weakly native small mammals and take on their fleas A change of habitat may be followed by an exchange of fleas.

Some interesting work has been done in testing the appetite of different kinds of flea for human blood. The oriental rat-flea (*X. cheopis*) has been kept alive for three weeks on that diet. Other species show repulsion for mankind and refuse to suck. The experiments confirm the popular belief that fleas have a marked preference for certain individuals. When the flea has refused to bite the human arm, it becomes necessary to check the experiment by trying whether the refusal is merely due to want of hunger. For this purpose a rat must be at hand. It can be secured on a board by two bandages fixed at each end by drawing pins. The rat lies, of course, on its back with its head comfortably supported by a little pillow of cotton

wool.  A portion of the rat's abdominal wall is left
exposed and shaved.  The flea, in an inverted test-
tube, can then be put on the hairless patch of the
abdomen and given an opportunity of biting, which
it may or may not accept.

When fleas are being collected from rats it has
been noticed that the true rat-fleas are usually on
the hind-quarters of the.host, whilst the mouse-flea
prefers the region of the head and neck.

As regards the tastes and habits of oriental rat-
fleas in the matter of food a long series of experiments may be summarized in this way :

(1)  When many rat-fleas are present some will
attack man, even when a rat is available for their
food-supply.  (2)  When the number of rat-fleas is
small, and when their true host is present, they will
not attack man.  (3)  When rat-fleas are starved
they will readily attack all animals, not being
particular in the choice of a host.  (4)  Rat-fleas
deprived of their food for from 72 to 96 hours
attack and feed on man more readily than at other
times.  (5)  Rat-fleas, even when starved, prefer
their true host to man.  (6)  Rat-fleas may be
attracted to man, jump on him, but take some time
to feed on him.  Plague-infected fleas might in this
way be carried from one place to another without
infecting the man ; but they would, when brought
near a rat, attack it in preference to man.

The fleas found on bats possess certain peculiarities which have led to their being grouped together. They form a family to which the name *Ceratopsyllidæ* has been given. They are recognized by two flaps, one on each side of the head. What these are and what service, if any, they render to their possessors is unknown. Bat-fleas also, as a rule, have maxillæ shaped like dumb-bells; but in one genus (*Thaumopsylla*), found on fruit-bats, they are triangular as in other fleas. The maxillæ, as the reader may remember, are parts of the insect's mouth, and, though placed like jaws on each side of the aperture, they are not used in piercing the skin and sucking blood (Fig. 4). They bear feelers called the maxillary palpi. The flea (*Thaumopsylla breviceps*) which is found on South African fruit-bats and which has triangular maxillæ, seems to be a connecting link between this peculiar group of fleas and the main family *Pulicidæ*.

Bat-fleas are commonly well supplied with combs. They usually have them on the abdomen, as well as the head, and the maximum number of eight combs is found in bat-fleas. Their structure and life-history agree generally with that of other fleas. They breed in hollow trees, caves, ruins, church-towers and lofts where bats hibernate or spend the hours of daylight. The larvæ feed on the droppings of the bats, and the mature insect, after emerging from the pupa case,

takes the first opportunity that comes of getting on
to its host. Bats are seldom found to be much
infested with fleas; for this reason, bat-fleas are
somewhat difficult to obtain and many of the species
that are known are extremely rare.

The hosts of bat-fleas, obviously, vary more as to
the surroundings which they inhabit than almost any
other animals. They are found from the equator
north and south to the Arctic circle and the straits
of Magellan, in the densest tropical forests and flitting
round the barest northern buildings. Some pillage
the rich fruit gardens of India, whilst other smaller
bats work hard for a precarious diet of gnats round
a Siberian village. Two sharply divided groups of
bats exist: (1) The fruit-bats (Macrochiroptera) with
flat molar teeth adapted for a vegetable diet. These
are found in the warmer parts of the Old World but
not in America. (2) The insectivorous bats (Micro-
chiroptera) whose molar teeth are equipped with
sharp cusps for biting their animal food. These have
an almost world-wide distribution, and one species at
least ranges within the Arctic circle. The same fleas
are not as a rule found on the large fruit-bats as on
the small ordinary bats. But some bat-fleas have an
extensive range. The same species has been taken
from different bats of various kinds in Sierra Leone,
in Madagascar and in Java.

All bat-fleas are blind. This absence of eyes, in

fleas which are parasites of strictly nocturnal animals,
lends colour to the suggestion that fleas which are
blind have lost their eyes because they had no need
of them. Disuse is speedily followed by degeneration.

# APPENDIX A

## SYSTEMATIC VIEW OF THE ORDER
### *SIPHONAPTERA*

Order SIPHONAPTERA. Latreille (1825).

Insects with body laterally compressed. Head
rounded and fixed by the whole posterior part to the
thorax. Mouth-parts for piercing and sucking, consisting of paired mandibles with serrate margins and
unpaired labrum. These are sheathed by the labium
and labial palpi. Maxillæ usually triangular with
four-jointed palpi. Eyes simple, placed in front of
the antennæ, occasionally rudimentary or absent.
Antennæ of three main segments which lie when at
rest in a groove. Three thoracic segments, always
free, each consisting of a notum and a sternum. The
sterna of the second and third segments are further
divided into a sternum, an episternum and an
epimeron, the two latter constituting the pleura.
Wings and rudiments of wings entirely absent.

Abdomen of ten segments of which the sternite of
the first segment is suppressed. Abdomen enormously
swollen in pregnant females of certain species. Combs
frequently present on head, thorax, and abdomen.
Legs developed for leaping. Coxa powerful; femur
thickened; tarsi of five segments, ending in two claws
on the distal segment. Metamorphosis complete.
Larva of thirteen segments. Pupa enveloped in silken
cocoon. Imago a temporary parasite (usually) on
warm-blooded vertebrates.

I. Family *Sarcopsyllidæ*. Taschenberg (1880).

Rostrum (= labium + labial palpi) rather long
but very weak and pale, consisting of two or three
segments inclusive of the unpaired basal segment.
Genal edge of head always produced downwards into
a triangular process situated behind the insertion of
the maxillæ at the ventral oral angle. Thoracical
tergites together shorter than first abdominal tergite.

To this group belong the chigoes and their allies,
the most truly parasitic fleas. About fourteen species
have been described, which can be grouped into three
genera, viz. *Dermatophilus, Hectopsylla, Echidnophaga.*

II. Family *Pulicidæ*. Taschenberg (1880).

Rostrum (= labium + labial palpi) more or less
strongly chitinized, consisting, except in a few cases,

of five, or more, segments inclusive of the unpaired
basal one. Thoracical tergites together longer than
first abdominal tergite.

Here belong the majority of *Siphonaptera*.

III. Family *Ceratopsyllidæ*. Baker (1905).

Head on each side with two flaps situated at the
front oral corner. Here belong the bat-fleas only.
There are several genera, and about twenty-five
species have been described. In most of the bat-
fleas the maxillæ are shaped like a dumb-bell, but
in the genus *Thaumopsylla* they are triangular as in
the *Pulicidæ*.

———

Oudemans (1909) has put forward an alternative
classification of the order *Siphonaptera* based on
the morphology of the head:—I. *Integricipita*,
II. *Fracticipita*.

## APPENDIX B

### A LIST OF BRITISH FLEAS AND THEIR HOSTS

A list of the British Fleas (*Siphonaptera*) revised
to March 1913 :

| Names | Usual Hosts |
|---|---|
| PULEX, *L.* | |
| P. irritans | Man, Badger |
| XENOPSYLLA, *Glink.* | |
| X. cheopis, *Rothsch.* | Rat |

| Names | Usual Hosts |
|---|---|
| ARCHÆOPSYLLA, *Dampf* | |
| A. erinacei, *Bouché* | Hedgehog |
| CTENOCEPHALUS, *Kolen* | |
| C. canis, *Curt.* | Dog |
| C. felis, *Bouché* | Cat |
| SPILOPSYLLUS, *Baker* | |
| S. cuniculi, *Dale* | Rabbit |
| ORNITHOPSYLLA, *Rothsch.* | |
| O. lætitiæ, *Rothsch.* | Puffin and Manx shearwater (Scilly Is. only) |
| CERATOPHYLLUS, *Kolen* | |
| C. fasciatus, *Bosc.* | Rat |
| C. londiniensis, *Rothsch.* | House-mouse, Rat |
| C. sciurorum, *Schrk.* | Squirrel, Dormouse |
| C. melis, *Wlk.* | Badger |
| C. mustelæ, *Wagner* | Bank-vole and Field-mice |
| C. penicilliger, *Grube* | Bank-vole and Field-mice |
| C. walkeri, *Rothsch.* | Stoats, Voles, and Field-mice |
| C. gallinæ, *Schrk.* | Chickens and many birds |
| C. fringillæ, *Wlk.* | Sparrow |
| C. garei, *Rothsch.* | Many birds |
| C. rusticus, *Wagner* | House-martin |
| C. farreni, *Rothsch.* | House-martin |
| C. hirundinis, *Curt.* | House-martin |
| C. columbæ, *Gerv.* | Pigeon |
| C. styx, *Rothsch.* | Sand-martin |
| C. gallinulæ, *Dale* | Many birds: especially fresh-water breeders |
| C. vagabundus, *Bokeman* | Nests of Sea-fowl |
| C. borealis, *Rothsch.* | Gannet and Rock-pipit |
| C. rothschildi, *Waterst.* | House-martin |
| CTENOPHTHALMUS, *Kolen* | |
| C. agyrtes, *Heller* | Field-mice and Voles |

| Names | Usual Hosts |
|---|---|
| CTENOPHTHALMUS, *Kolen* (cont.) | |
| C. agyrtes nobilis, *Rothsch.* | Water-rat |
| C. bisoctodentatus, *Kolen* | Mole |
| DORATOPSYLLA, *Jord. and Rothsch.* | |
| D. dasycnemus, *Rothsch.* | Shrew |
| RHADINOPSYLLA, *Jord. and Rothsch.* | |
| R. pentacanthus, *Rothsch.* | Weasel |
| R. isacanthus, *Rothsch.* | Bank-vole |
| PALAEOPSYLLA, *Wagner* | |
| P. sorecis, *Dale* | Shrew |
| P. minor, *Dale* | Mole |
| P. kohauti, *Dampf* | Mole |
| LEPTOPSYLLA, *Rothsch.* | |
| L. musculi, *Duges* | Mouse |
| L. spectabilis, *Rothsch.* | Bank-vole |
| TYPHLOCERAS, *Wagner* | |
| T. poppei, *Wagner* | Long-tailed Field-mouse |
| HYSTRICHOPSYLLA, *Taschb.* | |
| H. talpæ, *Curt.* | Mole |
| ISCHNOPSYLLA, *Westw.* | |
| I. elongatus, *Curt.* | Noctule Bat |
| I. intermedius, *Rothsch.* | Serotine Bat |
| I. simplex, *Rothsch.* | Natterer's Bat |
| I. octactenus, *Kolen* | Pipistrelle Bat |
| I. hexactenus, *Kolen* | Long-eared Bat |
| NYCTERIDOPSYLLA, *Oudemans* | |
| N. longiceps, *Rothsch.* | Pipistrelle Bat |
| N. eusarca major, *Rothsch.* | Noctule Bat |

# APPENDIX C

## ON COLLECTING AND PRESERVING FLEAS

There are two methods by which fleas may be preserved for study when they have been collected. The first is by keeping the specimens in small tubes of alcohol; the second is by mounting each in Canada balsam on a slide for the microscope. The advantage of the former method is that the material can be used for dissection. The student can do nothing without a microscope, though some of the commoner species can be identified with tolerable certainty by a practised eye which is assisted by a pocket-lens.

The tubes are best stored away in a cabinet fitted with wooden shelves and holes to take the tubes like a test-tube holder. Fleas dried and preserved loose in a box, or gummed on card, are useless for purposes of minute examination, and are soon destroyed.

Fleas may be collected from the great majority of mammals and birds in almost all parts of the globe. They can be found in the hair and under the feathers, and also in the places where the animals habitually sleep. The best places, from which a plentiful haul may often be obtained, are the holes and nests in which the young have been reared.

It is essential to remember, when an animal has

been killed, that all the fleas leave as soon as the
body of the host gets cold. No time, therefore, should
be lost in searching for specimens.

If the animal is small enough it may be put into
a cardboard box, or a white linen bag, and a few
drops of chloroform or benzine can be poured on it.
In a short time the fleas will be found dead in the
bag or at the bottom of the box. Some may also be
found in the hairs and feathers when they are turned
back.

In the case of a large mammal the hair must be
turned backwards shortly after death, when the live
fleas may be seen running about and caught. For
this purpose a small camel's hair brush is very useful.
If a flea is touched with a brush of this kind which
has been dipped in chloroform, benzine, or alcohol,
the insect sticks to the brush, but can be easily floated
off into the tube of preservative. The best preserving
liquid is 50 per cent. alcohol. Methylated spirits can
be used. Acetic acid can also be used; but it is
objectionable because in a short time it destroys the
corks of the tubes.

Each tube should only contain the fleas collected
from one host, but as many specimens as possible
should be secured, because there may be several
species of flea on the same host.

The tube must be securely corked and labelled,
with the date, the locality, and the name of the host.

In foreign countries it may not be always easy to do this. In such cases the skin of the host must be preserved with the tube for subsequent identification. A number corresponding to the label on the tube should be attached to the skin of the host.

Fleas collected without records of the host from which they were obtained are of little or no scientific value. For this reason a tube should contain the parasites of one host only.

A convenient way of preserving records temporarily is to write in *pencil* on a small piece of paper which can be rolled up and put in the alcohol in the tube.

Small mammals generally, including bats, are good hosts; Rodents and Insectivora afford usually the most fruitful captures. In trapping mice and voles only those traps should be used in which the animals are caught alive, or the fleas will have left their hosts before they can be secured and examined. Field-mice caught in the ordinary small penny mouse-trap are often found dead in the morning. The best traps are made on the principle of the ordinary mouse-trap, but larger. A piece of bacon-rind on the hook is a good bait for almost all small mammals. Where a number of traps are put down and left out they should, of course, be visited daily.

When a live mouse, or other small mammal of similar size, has been captured it may be transferred from the trap into a small white linen or holland bag.

The animal can then be killed by tapping its head or breaking its neck from the outside of the bag. After this has been done, fleas may be searched for in the fashion described above.

Most of the small mammals which act as hosts for fleas are nocturnal. The localities where they may be trapped are numberless, but only a small proportion of the captures may yield anything for the flea-collector. I have heard of a collector of small mammals who travelled through remote parts of Spain and never lost an occasion for putting down his traps when he had to change trains at a country railway junction. In England it would, however, seldom be worth doing this, as, on many lines, there is an attempt to make the arrival of one train and the departure of another correspond.

The following plan for securing bird-fleas will be found successful. The nests of birds should be taken as soon as the young are fledged and flown. If the nest is small it can be put into a glass-topped box lined with white paper. If too large, the whole nest, or the most firmly matted and dirty part, may be put into a glass globe (such as gold-fish are kept in) and a piece of paper tied over the mouth. From time to time the nest should be slightly damped with water. In every case a label should be put into the receptacle to preserve the name of the bird which built the nest. The bottom of the nest may sometimes be seen to be full of the larval fleas; but in any case fleas will probably emerge from pupæ. The fleas will continue appearing for as much as six weeks or eight weeks after the young birds have left the nest. They must be watched for and taken off the sides and top of the box with a camel's hair brush dipped in chloroform or benzine. I have seen dozens of fleas come from the nest of a tit (*Parus major*) in the course of a few weeks. They were all of the common species (*C. gallinæ*). When the lid of the box is removed it is difficult to prevent a few escaping in the room, but I have never known them cause inconvenience to anyone.

Collectors in warm countries should give their attention to the chigoes and their allies, which are of great interest and have been little studied. They are found on mammals and birds in tropical and semi-tropical countries. The males are very difficult to find, but the females are large and very parasitic. They have the appearance of a small wart firmly fixed to the skin. Small mammals may be transferred, with their chigoes attached, to a bottle of alcohol. Many examples of these insects are often found together on the more naked portions of their hosts.

# APPENDIX D

## BIBLIOGRAPHY

A. General Bibliography of the chief and most recent works. Many have bibliographies of earlier works.

1. TASCHENBERG (1880), *Die Flöhe.* Halle.

    Although the author only describes 30 species, his book forms the basis of all subsequent scientific work. He divides fleas into (*a*) chigoes, (*b*) non-chigoes. Plates. Bibliography. References to 73 earlier writers.

2. KARSTEN (1864), *Beitrag zur Kenntniss des Rhyncoprion penetrans.* Bull. Soc. Imp., Moscou. Vol. XXXVII., p. 72.

    A full account of the life and habits of the chigoe. Many references to older writers and travellers.

3. WAGNER (1889), *Aphanipterologische Studien.* Hor. Soc. Ent. Ross. Vol. XXIII., p. 199 ; (1893) Vol. XXVII., p. 347 ; (1898) Vol. XXXI., p. 539 ; (1902) Vol. XXXV., p. 17 ; (1903) Vol. XXXVI., p. 125.

    A series of learned papers by a professor at the Russian University of Kieff. He has also written in Russian.

4. DAMPF (1907), *Die Ost- und Westpreussische Flohfauna.* Schriften der Physik.-ökonom. Gesellschaft zu Königsberg-i.-Pr., XLVIII. Jahrgang, p. 388.

    Contains an excellent general account of our present knowledge of fleas. Short Bibliography.

5. DAMPF (1910), *Palaeopsylla Klebsiana n. sp. ein fossiler Floh aus dem baltischen Bernstein.* Schriften der Physik.-ökonom. Gesellschaft zu Königsberg-i.-Pr., LI. Jahrgang, II. 248.

    Good plates of the fossil flea in amber.

6. OUDEMANS (1909), *Neue Ansichten über die Morphologie des Flohkopfes sowie über die Ontogenie, Phylogenie und Systematik der Flöhe.* Novitates Zoologicæ, Vol. 16, p. 133.

    Suggests a new classification based on the morphology of the head, viz. (1) Integricipita; (2) Fracticipita. By the chief authority on fleas in the Netherlands. He has also written numerous papers in Dutch.

7. JORDAN and ROTHSCHILD (1908), *Revision of the non-combed eyed Siphonaptera.* Parasitology, Vol. I., p. 1. Plates. Bibliography.

    An excellent piece of work, which includes an account of the plague fleas.

8. JORDAN and ROTHSCHILD (1906), *A Revision of the Sarcopsyllidæ.* Thompson Yates and Johnston Laboratories Reports, Vol. VII., p. 15. Plates. Bibliography.

    The best modern account of the chigoes.

9. BAKER (1904), *A revision of American Siphonaptera or Fleas together with a complete list and bibliography of the group.* Proc. U.S. National Museum, Vol. XXVII., p. 365. Plates. Bibliography of special papers only.

    The earlier references, beginning 1699, mostly from Taschenberg.

10. BAKER (1905), *The Classification of the American Siphonaptera.* Proc. U.S. National Museum, Vol. XXIX., p. 121.

    Gives an index of hosts and their fleas. Additional bibliography.

11. ROTHSCHILD (1898), *Contributions to the knowledge of the Siphonaptera.* Novitates Zoologicæ, Vol. 5, p. 533; (1900) *Further Contributions, etc.,* Vol. 7, p. 539; (1903) Vol. 10, p. 317; (1904) Vol. 11, p. 602; (1905) Vol. 12, pp. 153 and

479; (1907) Vol. 14, p. 329; (1909), Vol. 16, pp. 53, 57, 61, and 332; (1906) *Notes on Bat-fleas*, Vol. 13, p. 186.

These papers contain, for the most part, descriptions of new species in the writer's collection. Many fine plates illustrating morphology and structure of the external skeleton.

B. The following references are to papers on the systematic position of the *Siphonaptera* and their relationship to other insects.

1. KRAEPELIN (1884), *Ueber die systematische Stellung der Puliciden.* Hamburg, 1884.

2. PACKARD (1894), *The Systematic Position of the Siphonaptera.* Proc. Boston Nat. Hist. Society, Vol. XXVI., p. 312.

3. DAHL (1897), *Puliciphora, eine neue flohänliche Fliegengattung.* Zoologischer Anzeiger, Vol. XX., p. 409.

4. WANDALLECK (1898), *Ist die Phylogenese der Aphaniptera entdeckt?* Zoologischer Anzeiger, Vol. XXI., p. 180.
   A humorous reply to Dahl.

5. DAHL (1899), *Die Stellung der Puliciden im System.* Archiv für Naturgeschichte, Vol. 65, I. p. 71. Plates.

6. HEYMONS (1899), *Die systematische Stellung der Puliciden.* Zoologischer Anzeiger, Vol. XXII., pp. 223 and 301. Three figures.
   A destructive criticism of the views advanced by Dahl.

7. SEMENOV (1904), *Zur Frage der systematischen Stellung der Flöhe.* Revue Russe d'Entomologie, Vol. IV., p. 277. In Russian.

C. The following references are chiefly to works on plague and fleas.

1. ADVISORY COMMITTEE (1905–1909), *Reports on Plague Investigations in India.* Journal of Hygiene, Vols. V., VI., VII., VIII. and X.
   These volumes contain the five "Extra Plague Numbers." Many references to observations and experiments on rats and fleas.

2. HANKIN (1905), *Plague Epidemiology.* Journal of Hygiene, Vol. V., p. 48.

3. OGATA (1897), *Ueber die Pestepidemie in Formosa.* Centralbl. für Bacteriol., Vol. XXI., p. 769.

4. SIMOND (1898), *La Propagation de la Peste.* Annales de l'Institut Pasteur. Vol. XII., p. 625.

5. HANKIN (1898), *La Propagation de la Peste.* Annales de l'Institut Pasteur, Vol. XII., p. 705.

6. VERJBITSKI (1908), *The part played by insects in the epidemiology of the plague.* Journal of Hygiene, Vol. VIII., p. 162.
   Experiments made 1902–3 at Cronstadt and S. Petersburg with fleas. This important research was written in Russian and not translated or published till 1908.

7. DOANE (1910), *Insects and Disease.* Constable and Co.
   A popular work by an American entomologist. Contains a chapter on fleas and plague. Some good micro-photographs of fleas. Bibliography.

# INDEX

For EU product safety concerns, contact us at Calle de José Abascal, 56–1°,
28003 Madrid, Spain or eugpsr@cambridge.org.

www.ingramcontent.com/pod-product-compliance
Ingram Content Group UK Ltd.
Pitfield, Milton Keynes, MK11 3LW, UK
UKHW010851090126
466816UK00011B/156